SEA

A User's Guide

Sonia Surey-Gent and Gordon Morris

Whittet Books

First published 1987
This paperback edition published 2000
Reprinted 2004
Text © 1987 by Sonia Surey-Gent and Gordon Morris
Illustrations © 1987 by Sonia Surey-Gent

Whittet Books Ltd, Hill Farm, Stonham Road, Cotton,
Stowmarket, Suffolk IP14 4RQ
Distributed in Canada and USA by Diamond Farm Book
Publishers, PO Box 537, Alexandria Bay, NY 13607

Design by Richard Kelly

ISBN 1 873580 54 1

Printed and bound in Malta by Interprint Limited

Contents

Foreword

This present book, intended as a practical user's handbook, was written at the instigation of its publisher, Annabel Whittet, well known for her vegetarian restaurant guide. She turned to seaweed as a means of increasing the variety available to vegetarians, and first of all approached John Seymour on this matter. He is, of course, the leading proponent of self-sufficiency in this country, and in small-boat circles is also known for his coastal voyages in the coble *Willy-Nilly* and other seafaring adventures. John was kind enough to recall that I had been in contact with him some time previously concerning the uses of seaweed as a soil conditioner, one aspect of my wider interest in the whole subject. I in turn passed on the notes of my researches to Sonia Surey-Gent, a marine biologist. These have been greatly extended by her knowledge and experimentation to produce this book. In particular, all recipes are the product of her cooking skill and ingenuity. Even where they are inspired by traditional or foreign dishes, they have been reformulated by her to suit source materials available in Great Britain, and worked out to suit cooking in the modern kitchen.

Seaweed, its role when living and its derivations, provide such a wide field of interest that it has been difficult to select items for coverage in this book. I hope that we will have succeeded if we help to stimulate readers to the useful and interesting aspects of their environment of which seaweed forms such an essential and neglected part. It is of the utmost importance that as many folk as possible realize how much is at stake in the conservation of the delicate life-system which supports all other life on this planet. Then politicians can be pressurized to take adequate measures to sustain it; whereas now there are cut-backs in expenditure on organizations directed to the understanding of this marine environment.

As a member of the Marine Biological Association (U.K.), I wish to acknowledge the role of the association, with its excellent specialist library facilities for members. Here I have spent many happy hours among the proceedings of international symposia and similar specialist documents. Also I am most grateful for encouragement from Dr Boalch of the Senior Scientific Staff of the M.B.A.

Having spent a lifetime upon practical works, largely concerned with marine regimes, including the installation of marine outfalls and the organization of marine dumping, I can appreciate the magnitude of pressure which human technology is now exerting on the natural ecosystem of the sea and its margins and have become increasingly concerned about it. In stating the case for marine conservation one cannot separate it from the whole cause of environmental conservation of Planet Earth. It has been suggested that a better name would be 'Planet Sea' because of the great preponderance of oceans upon the earth's surface: not only do oceans take up more of the surface area but they are extremely important as originators and sustainers of its atmosphere and all dependent life. This role as 'keeper' of the atmosphere is rendered even more important in view of the current trend towards destruction of the rain-forests. This basic vital role of the sea, and principally of its coastal margins, is performed by the flora, marine algae or seaweeds which form the major part of plant life on this planet.

The whole structure and interaction of life in the sea is a complex one; it cannot be emphasized too often that the entire ecosystem must be preserved if mankind is to continue living on this blue planet. In *The Gaia Atlas of Planet Management* is a collection of contributions from specialists in various planetary ecosystems, developed from a theme first proposed in Dr James Lovelock's book *Gaia: A New Look at Life on Earth*. He proposed the hypothesis that the biosphere of earth has become organized as an interlocking, self-perpetuating, life-support system. Dr Lovelock discovered that some seaweeds cause iodine to be released into the atmosphere from the sea, in the form of its methyl compound. This atmospheric source is essential for both animal and plant growth on land. Knowledge of the marine ecosystem is still incomplete, but this should not be the excuse to delay essential conservation. Likewise, the long-term effects of persistent low-level pollution are not fully understood, except that they are bad, and action is needed. Governments persist in regulating only discharge rates, while ignoring the more important total discharge of pollutants.

In conclusion, I would like to thank James Wharram, pioneer catamaran designer and editor of the magazine *Sea People*, for his encouragement, and the use of part of his stand at the Earls Court Boat Show several years ago. This enabled me to show off my ideas for the use of Polynesian-type catamarans as bases for floating seaweed culture and sea-fish farming. As a direct result of this show I met Sonia Surey-Gent, and finally a book has emerged. Thanks also to Sydney Surey-Gent, her long-suffering husband, who has been the recipient of seaweed dinners, sometimes unknowingly, and who has been heard to utter appropriate comments on the dishes, not always bad.

Gordon Morris
February 1987

Introduction

To most people seaweed is that frondy stuff that wraps around their ankles when paddling in the sea, or the slippery blanket over rocks and pools when the tide goes out. Many will recognize the broad-bladed kelp that is supposed to forecast the weather, and the pop-weed that crackles underfoot. But few understand the importance of these seaweeds on a planet where more than two-thirds of its surface is covered by water. Many people don't like the word 'weed' so refer to macro-algae, the large seaweeds, as sea-vegetables, but they are far more useful than that name implies. The change in modern awareness of conservation issues, self-sufficiency, wild foods and re-found knowledge means that 'weed' is now acceptable, an honest word for an honest plant. So we make no apologies for using 'weed' of seaweed as a name, or for dinner.

Of the thousands of different types (or species) of seaweeds, there are more than eight hundred species around Britain alone. These play a very important part in renewing the air we breathe and keeping the borders of the ocean the thriving balanced communities they are at present. If you could weigh the mass of vegetation that fills the shallows of the world's seas you would find that it exceeded that of all the world's land plants. This last great natural resource of our planet is still largely unexploited by man. Yet in an age of ecological awareness, as well as rising unemployment, surely the harvesting of a product that is self-renewing and needs only manpower to gather it should catch the interest of the adventurous businessman.

Macro-algae are useful to man in many ways. Other countries have used the plants whole, or products from them, for centuries. In the UK there have been a few coastal areas and islands where people have harvested seaweeds, but never to the extent found in Japan. However, modern methods of food processing and other technological changes have brought seaweed products into everyday life, without people realizing what it is they are eating or using. Seaweeds as health foods are now becoming more popular, especially now that western society is discovering that it is good to be healthy. It is a shame that fashion has to rule before something so vital to a good life should become easily accessible in the shops. Dried seaweed is imported from Japan via the USA and from Ireland via Scotland. Like yoghurt, once considered a 'crank' food and now universally popular, seaweed is probably destined to become the gourmet dish of the future. Meanwhile, beside the dedicated experimenters with anything new, lovers of Japanese cuisine and writers of books on seaweed, the general public remains mostly oblivious to weedy dishes and preparations. The chapter on commercial uses will prove an eye-opener for many. Body builders and athletes consume kelp tablets as aids to their training and diet regimens; slimmers are discovering *Spirulina*; rheumatism sufferers bathe in seaweed extract. Old remedies for old and new problems, but, if your doctor gave you a diet of seaweed and told you to be patient instead of a box of pills with a promise of instant relief plus side effects, which would you choose? Once despised old-fashioned ideas of what is natural and organic are now becoming 'the latest thing'. People are like plants. Change does not come instantly, you must grow into change. Learn to overcome simple ailments yourself. Let super-science do those things it can do so very well, like microsurgery that can save a severed hand, or replace a malformed kidney. As for aspirin, you can make your own . . . Salicin is found in willow bark. In the body it is converted to salicylic acid, which is closely related to aspirin. The bark can be infused in water or taken as a powder.

The authors of this book want you, the reader, to *use* seaweed. Use it in the kitchen, on the garden, to feed animals, to make medicine and cosmetics. You do not need highly specialized knowledge or techniques, and the raw materials come free. A few books have been written about sea-vegetables, mainly recipe books, mostly by American authors. This book tries to give some idea of the variety of uses seaweed has, not just for food. Put the book in your pocket when you go to the seaside. *Use it.* Remember, until you have tried seaweed for yourself, you cannot honestly give an opinion on it. The Bibliography at the end of the book will provide you with access to even more facts and figures about this wonderful wet world of plants. There are some unique recipes in the appropriate section that have never, as far as we know, been printed in any recipe book to date. Many are part of the unwritten folklore of the British Isles. The recipes are mainly vegetarian. Seaweed is a more natural food to man than is red meat. You can always adjust a recipe, because no two people ever agree on what is the best taste or texture. So don't be afraid to experiment.

Research continues, and new uses of seaweed are always cropping up, though there is often controversy when two scientific minds cannot agree. We have tried to present a balanced view of these disagreements and used the practical person's approach when presenting conclusions. If you have any ideas of your own, we would be glad to hear them, for, to quote D.T.V. Desikachery, a leading phycologist (seaweed scientist), writing in *Marine Plants*: 'plants and animals abound (in the sea) in such opulence that it is not possible for any single person to even describe all of them. Marine plants by themselves form the very basis for all other forms of life to exist or continue to survive.' We have not tried to describe more than a few of the commonest seaweeds, with the view of actually turning them into food. The one baddie in the bunch, *Desmarestia* species, is also detailed, though you are unlikely to come across it unless you are a diver. Unlike the mushroom and toadstool groups, there are no poisonous seaweeds, some just don't taste so good. Others are delicious, so *bon appétit*.

Sonia Surey-Gent
February 1987

1

History

Imagine yourself back in the Stone Age, down by the sea. Early man is away shouting the odds at a neighbour, disputing territory or trying to catch something meaty that can outrun him seven times over. Early woman meanwhile gets on with the serious job of providing everyday food. She is used to the water, wary only of big fish with sharp teeth or stingers in the sand. The tide goes out, revealing colourful plants and shellfish, and soon she has a pile of tasty bites big enough to keep a starving brontosaurus happy, if one had been around then. When early man returns home empty-handed, he tucks into a meal of sea foods, and leaves the remains in his rubbish tips for us to find in the twentieth century.

Prehistoric records of seaweed eating, such as that found in the coastal region of South Africa, are few and far between, but common sense would dictate that an easily gathered tasty food from the sea would figure more often in the diet of shore dwellers than a hard-won carcass. Earliest records from 3000 BC indicate that in China the emperor Shen-neng used seaweeds as both medicine and food. In 800 BC the *Chinese Book of Songs* mentions a housewife cooking sea vegetables, and in the *Erh Ya* from 300 BC, the oldest known Chinese encyclopaedia, twelve species of seaweed are recorded. Even the Koreans used to send sea plants to the Imperial Court of China, these being listed as kelp, agar and nori. There would have been a good trade along the coast and inland across China, and it is reasonable to assume that everyone enjoyed the benefits, because the weeds turn up in what would be considered simple peasant recipes, traditional to inland provinces of the Chinese mainland today.

The Japanese loved not only the taste, texture and benefits of seaweed, but also the grace and beauty of the plants themselves, and wrote many poems about or referring to the weeds. It was a compliment for a lady to be likened to the pliant, supple sea plant. All the main seaweeds that are consumed in Japan today are listed as annual tribute to the court in the eighth century.

Back in the West, Iceland has used seaweed since the first Vikings landed there. It has provided food for both men and beasts, been used as fuel (the ash is good in the garden), as manure and mulch and to protect plants from the frost. Dried weed was also stuffed into cushions, and strands used as lampwicks. As a vegetable it provided all the vitamins and minerals needed, plus a cure for seasickness, and finally a dye used on homespun woollen materials. In fact, seaweed gathering was so important that a law had to be made setting out the rights and concessions involved before anyone could collect and/or consume fresh 'sol' (*Rhodymenia palmata*) on a neighbour's land.

Pythagoras, the Aztecs, Indians and Vikings all recorded the eating of seaweed. Hawaiians used to cultivate 'limu', seaweed gardens of over seventy species, until western eating habits corrupted their good taste. Now the knowledge is lost, and, too late, a renewed consciousness of their own history has set Hawaiians to revive the limu garden. In Britain, coastal dwelling people ate seaweed but only a few instances have been recorded for posterity: the Romans noted that the British in Wales ate water plant or laver, others used carragheen and dulse. As late as the 1800s young stalks of sugar wrack (*Laminaria saccharina*) were sold on the streets of Edinburgh.

All the seaweed used until modern times was gathered by hand: a labour-intensive, seasonal occupation, with products meant only for local consumption in most areas. Wave-washed sea wrack was hauled up onto the fields to improve soils, from sand to heavy clay. Many lanes in the west of Britain called 'Donkey Lanes' were used by laden animals hauling seaweed and sand up the extremely steep paths from the beaches to the fields. Lanes were sometimes cut out of the cliff face, even stepped. The plants were used for food, both human and animal. Romans fed their horses on a ration of seaweed with other foods. Latin ladies used seaweed to produce cosmetics, others to dye clothes.

In France in the twelfth century, deposits of a calcareous red seaweed called 'maerl' were used as fertilizer, indeed they still are. Channel Islanders spread seaweed on the ground, then ploughed it in. Recorded in the fifteenth century by a traveller, this practice was followed until recently by a few farmers on the islands. Gathering the storm wrack for the fields and gardens was a great social event in Ireland in the 1840s.

Commercialization began to creep in during the 1800s. In

1881 E.C. Stanford of Scotland patented his discovery of alginic acid, derived from sea algae. Not much happened until 1934 when development began and alginates were produced for use in both food and industry. Just prior to this, seaweed was being sold as a fertilizer and for stock feeding, with a commercial production plant being set up in Scotland in 1950. Mr W.A. Stephenson started the making and marketing of the liquid seaweed 'Maxicrop' based on experiments carried out by Dr Milton in 1944. The product is still going strong in the 1980s and has done much to pave the way towards the 'organic' growing of market garden produce.

Burned seaweed was used in the glass, soap and later iodine industries. In the Middle Ages, seawood was burned to produce salt, and France was the last country to use this old method, records ending in 1950. As a source of minerals and salts, seaweed is inevitably supplanted by easier, cheaper chemical methods based on land deposits of raw materials – for example the vast quantities of nitrates discovered in Chile in 1840 that ended seaweed burning for most countries. Unfortunately, anything grown in water is difficult to collect, sometimes unpleasant, and heavy to move. However, land-based deposits are not infinite and one day we may be very grateful for the provisions of seaweed.

Modern uses are discussed more fully in the commercial uses chapter (10). All over the world hand gathering of seaweed for home consumption still continues; there are no figures for such activities, very few records kept, but the sea continues to yield its harvest.

2

Identification

Classification

The algae are a group which contain some of the most primitive members of the plant kingdom, simple structures that appeared on earth very early in geological history and haven't changed very much to the present day. Large algae (macro-algae) are the seaweeds we are all familiar with. There are other types, microscopic single-celled varieties that make up the floating population of phytoplankton (*phyto* – plant, *plankton* – drifting or floating). These tiny plants are the ultimate basis of life, the bottom and beginning of every food chain. Of this latter group we are only concerned with a variety called *Spirulina*, for direct human food use. The common names of 'wrack' and 'kelp' found in many books cover whole groups of different species. All the *Fucoides*, such as *Fucus vesiculosus* and *Ascophyllum nodosum*, are lumped together under the general heading of 'sea-wrack' or 'wrack'. Kelp refers to the many species of *Laminaria*, but is most commonly used for the commercially harvested *Macrocystis*. Seaweeds are divided into three main groups, roughly distinguished by colour: *Chlorophyceae*, the green algae; *Phaeophyceae*, the brown algae; *Rhodophyceae*, the red algae, and a fourth group *Cyanophyceae*, the blue-green algae of which *Spirulina* is a member.

The colour of the plants in each group is more or less definitive. Remember that all have chlorophyll as their photosynthetic base, so all are green, but the green is then covered by another pigment, such as brown or red. Some colours may seem to overlap into another category. Shade can change with season, or age of the weed. Nevertheless colour is a useful guide to identification, and the light-absorbing characteristics decide where in the zonation scheme of things the seaweed is able to grow.

Chlorophyta (green) absorb the long wavelengths of light – red light – and reflect green. The plants grow where they are exposed to the strongest light, in shallow waters, and store food as starch,

15

just like land plants. Indeed they are linked with land plants, being on the borderline between land and sea. Pigments present are phycocyanin (blue), xanthophylls (yellows) and carotene.

Phaeophyta (brown) are the largest and longest of the seaweeds. They absorb medium wavelength green light and contain the pigment fucoxanthin. Green light can penetrate further into the water than can the red, so enabling brown weeds to live deeper. They also show a preference for cooler water temperatures.

Rhodophyta (red) seaweeds absorb the blue and ultra violet light wavelengths. Red is the most common colour of seaweeds, and they can live down to great depths – 600 metres is common in very clear water areas. Their main pigments are phycoerythrin (red) and phycocyanin (blue).

Cyanophyta contain the same pigments as the green seaweeds.

Algae are very simple plants made up of unspecialized cells that show no division into root, stem and leaf. They do not flower or fruit and have no need for special 'circulation' systems to carry food around the plant body. Seawater contains all the food that the weed needs to absorb, and at the same time gives support to the plant body. No rigid cellulose skeleton is required as in land plants, so prolonged cooking to break down the cellulose is unnecessary. Another difference from land-bound plants is that the seaweed grows from the base not the tip, so in any one frond of weed the youngest parts are at the bottom and the oldest parts at the top. The proper names of the parts of a seaweed, which you will find in books on the subject and in the next few pages of descriptions, are shown on the diagram.

Holdfast
This is root- or disc-like in appearance and is in fact no more than an attachment organ. It does not take up nourishment as do roots of land plants. Its main task is, as its name suggests, to hold the seaweed fast to the rock or stone surface. Sometimes the holdfast may creep over an area, spreading the plant over surrounding rock or shell surfaces. The larger holdfasts are favourite hiding places for many small sea creatures such as brittle-stars and worms.

Frond
Main body of the seaweed. It can vary in form from a flat blade

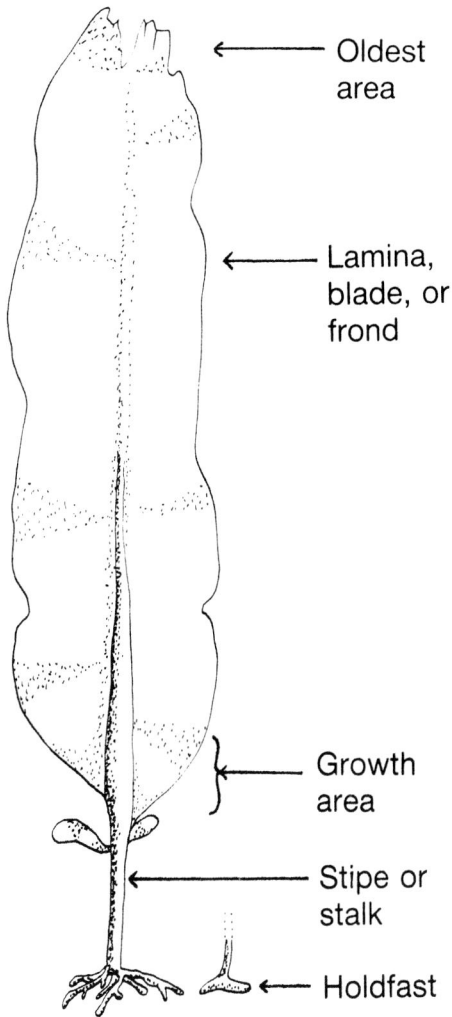

Oldest area

Lamina, blade, or frond

Growth area

Stipe or stalk

Holdfast

that looks like a conventional leaf to fine filaments that resemble hair. It may be one long single piece or branched into simple or complex shapes. At the base there may or may not be a stalk-like attachment to the holdfast.

Thallus
The whole seaweed plant. Its special characteristic is that, unlike a land plant, each cell in the seaweed carries out the same function as any other cell.

17

Seaweeds are classified as belonging to the plant sub-kingdom *Thallophyta*, in which there are eight classes altogether, but in this book we are concerned with only the large weeds (macro-algae) and one species in the micro-algae. These are identified primarily by colour and are as follows:

Chlorophyceae – green algae (macro)
Phaeophyceae – brown algae (macro)
Rhodophyceae – red algae (macro)
Cyanophyceae – blue-green algae (micro)

Reproduction
Seaweeds don't have flowers, fruits or seeds. Instead they produce spores which can swim, from special parts of the frond called sporangia. These spores are asexual, that is they do not take part in any sexual exchange with another spore, and when they settle onto a suitable piece of rock new plants will grow into either male or female weeds. Females then develop egg cells which remain fixed in the frond, and males develop swimming sperm cells. Both eggs and sperms are released into the sea, where fertilization takes place. Fertilized eggs then stick onto surfaces where sea conditions are favourable to the particular seaweed species, and grow into new plants. In some seaweeds the male and female plants are identical, other species may show colour differences. Further variations are introduced when the sexual form of a plant may be so different in shape from its asexual form that the two variants are identified as two completely different species. The alternation between sexual and asexual phases is known as 'alternation of generations' and is very much a seaweed speciality.

Just to add more confusion to the picture, not all seaweed species follow this general pattern, some have very complex variations, others have merged reproductive activities, and a lot is yet unknown.

Common names
There are so many common names for the same seaweeds in different parts of the country and over the world that it is safer to stick to the scientific names when dealing with identification. Many different Japanese weeds are given the same name because it describes their food value and eating quality, not the individual species, hence you will find 'kombu' and 'nori' labels on a variety of species. Other group names are 'wracks' – referring to the

Fucacea (bladderwrack, etc) – 'oar weeds' and 'kelps' – mainly *Laminaria* species – and sea-greens – the green weeds of all shades.

Proper names

All the seaweeds have two names; the first name (with an initial capital letter) is rather like our surname, it tells you the family or genus to which the seaweed belongs; within this genus there can be many 'children', or species, all bearing the same first name. The second name, which always starts with a small letter, not a capital, is the plant's own species name, just as all the children in one family have their own names. In the following identification lists, you will notice that after each plant's name there is often another one or two. These refer to the scientists who first named the seaweed. If there has been a revision or re-identification of that plant, then the first scientist's name will be in brackets, followed by the latest person's name.

For example: in the class *Chlorophyceae* (green seaweeds): *Enteromorpha intestinalis* (Linneaus) Link, *Enteromorpha* is the genus, *intestinalis* is the species; Linneaus first named this seaweed, but Link has subsequently revised or re-identified the species. This happens all the time in plant studies, and names are changed as the plant is put first into one genus then into another. The name is often a description of some characteristic of the plant – for example, *Enteromorpha intestinalis* is supposed to look like a portion of intestine, sausage-like with gas in it. If you like to follow the name game you will need a Latin dictionary and sometimes a bit of imagination, because there were occasions when the scientists had a bit of fun. Other seaweeds were named after their discoverer or someone who was important at the time. Sometimes it was a way of saying thank you to a benefactor who had perhaps funded an expedition.

The seaweeds included in this identification chapter are the most common and most abundant weeds, for each of which there is at least one recipe in the appropriate chapter. Uses of the plant as human food are briefly mentioned, other uses as animal food and in agriculture are so general that you are referred to the appropriate chapters for further information.

CHLOROPHYCEAE (Green seaweeds)

Chaetomorpha linum (C.F. Muller)
Green hair-weed

Frond Fine hair-like threads, unbranched and tapering towards the base. Each hair can be 40cm (16 inches) long or more, but they are usually so tangled together that the effect is of a green mat. If you use a hand lens, always an essential item of any shore browser's equipment, you will notice that the frond is made up of a single chain of large cells. This has been indicated on part of the diagram.

Holdfast No distinguishable holdfast, just a mat-like extension of the matted fronds, really modified cells at the base of each strand which 'stick' to the rock.

Colour Bright green, sometimes with a bluish tinge. Old plants may begin to yellow.

Texture Slippery yet somehow crisp to the touch.

Similars There are many strandy green weeds, but others may feel wiry, or have definite holdfasts. If you are uncertain, but the weed feels soft, not like wire, then use as if it were *Chaetomorpha*.

Habitat Low salinity can be tolerated, and it loves plenty of light, so *Chaetomorpha* is found in the upper and middle shore pools, tangled around any other weed that grows there. Sometimes sand seems to sprout green-hair weed as it grows up from buried rocks. The weed is occasionally free floating although experts think that such patches may be another seaweed called *Rhizoclonium*.

Distribution Very common all round the Mediterranean, up the Atlantic coast and as far as the Baltic.

Abundance Very common everywhere. Because it doesn't mind lower salinity it will often grow where fresh water is seeping out of cliffs or across beaches, diluting the seawater in pools so that other weeds cannot thrive.

Picking time   Gather fresh plants in spring, when their vitamin content is at its highest.

Food rating   Good. High in iron and sugar, vitamin C in spring.

Uses   Fresh as a salad item. Dried as a condiment.

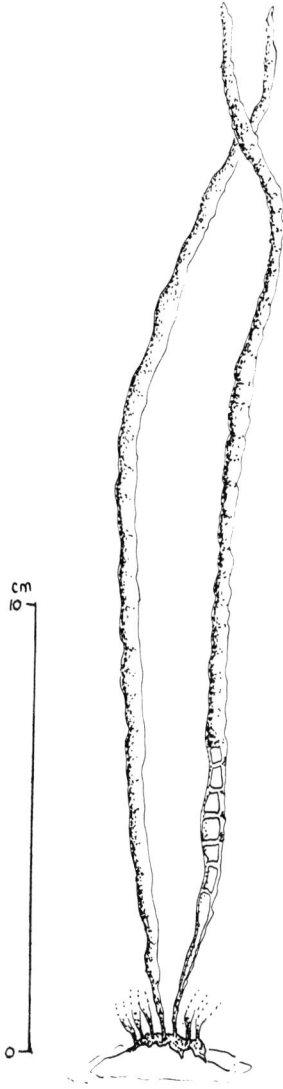

Codium tomentosum Stackhouse
Velvet horn, miru (Japanese for Codium fragile)

Frond The largest of the green seaweeds; a frond may reach 40cm (16 inches) or more where conditions are favourable. Typically it appears as a tubular, branching plant, each branch divided into two, then two again, known as dichotomous branching. Each branch may be about 1 cm (½ inch) in diameter, with rounded ends. The surface of the plant looks as if it is covered in green felt, like a snooker table.

Holdfast A bulky-looking disc that is in fact made of matted threads, covering the rock surface.

Colour Rich dark green. A singularly beautiful plant when in prime condition.

Texture Like velvet or felt all over the surface of the plant. May feel water-logged.

Similars Codium fragile is sturdier than its sister C. tomentosum, and may be found around British shores, but it is really a visitor from America.

Habitat From the middle shore down to the low spring tide mark, in pools hanging down from the rocks. Large clumps can form. Because this green weed does not have a spore stage, there is no alternation of generations.

Distribution From the Mediterranean to the English Channel, but it does not like water that is too cold or not salty enough.

Abundance Probably common, but records are rather sparse. Where it has been found it is very abundant, and can yield a good harvest.

Picking time Codium is a perennial plant which reaches its maximum growth in the winter. April and May are the best time to gather weeds, always leaving some of the clump intact, and never pulling off the holdfast.

Food rating   Good. The vitamin A content is said to match that of good cabbage, without the digestive problems that cabbage can bring.

Uses   It is popular as a vegetable, and is sometimes used as a condiment.

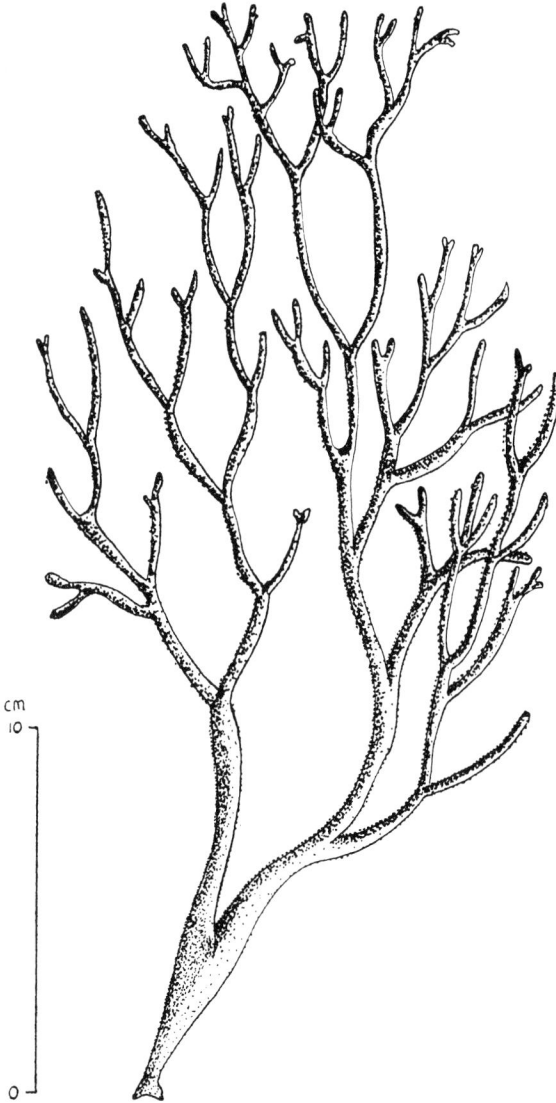

Enteromorpha species of many similar forms
Stone hair, green nori

Enteromorpha compressa Greville
Enteromorpha intestinalis (Linnaeus) Link
Enteromorpha linza (Linnaeus) J. Agardh, used to be *Ulva linza*

Frond   In all species the frond is long and thin. In *E. intestinalis* the frond can reach one metre (3 feet) long, its uneven outline being due to gas-inflated portions along its length. *E. compressa* is a thinner frond with branches and some authorities consider it to be just a variation of the first species. A hollow strand with crinkled edges is probably *E. linza*, and sometimes the edge may be extended to form a thin wavy sheet. There are many more species very similar, but the minute differences of each weed are really only of importance to taxonomists and biologists.

Holdfast   A tiny stalk that seems to stick into the rock surface. Sometimes there may be a fine mat of threads.

Colour   All shades of good leaf green from dark to pale. White along the edges shows where mature fronds have been releasing spores. All the plants look delicate and transparent.

Texture   A fine, smooth and slippery weed that can make rock scrambling dangerous for the unwary.

Similars   All the other *Enteromorphae*, and *Chaetomorpha* as well. *Enteromorpha clathrata* is the American variety.

Habitat   All the *Enteromorphae* are rock dwellers in the upper, middle and lower shore, in pools and around outflows, on the bottoms of boats, mooring ropes, and anything else that is left in the water for a while. They like places where fresh water mixes with salt water, especially if sewage is included. So check carefully before gathering for food, and, when in doubt, leave it (see p. 111).

Distribution   This family of weeds is very common, from the Mediterranean to the Baltic.

Abundance Too much of a good thing is usually the complaint of anyone who has fought their way across a meadow of this weed to reach the more interesting rock pools on the lower shore.

Picking time Early spring is the best time, especially if you are going to eat it fresh.

Food rating Very good. It is low on iron but well up on the trace elements.

Uses Along with sea lettuce, this is the popular 'green nori' of Japan, eaten raw, cooked or dried.

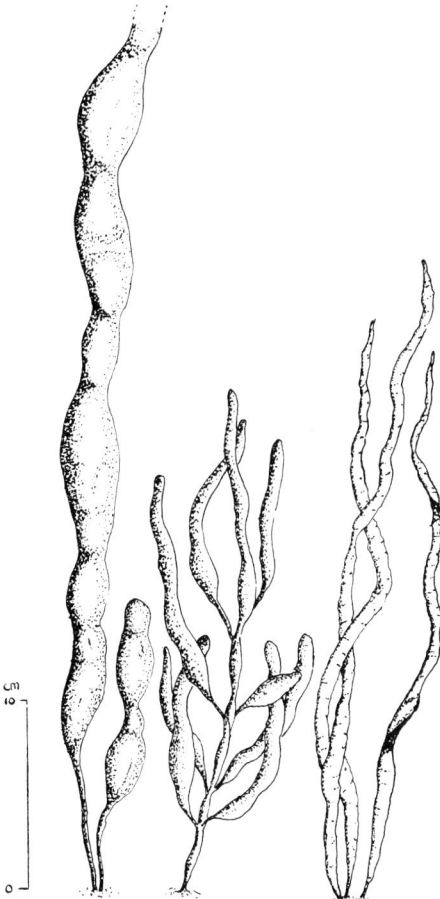

Ulva lactuca Linneaus
Sea lettuce, green laver, green nori

Frond   A broad-bladed, wavy-edged 'lettuce leaf', sometimes with a side lobe, or holes in the frond. Alternatively the whole frond may be elongated with a somewhat crinkled edge, but it is never ribbon-like as in *Enteromorpha linza*. If growing conditions are good, the leaf may reach 50 cm (20 inches) long or more.

Holdfast   A small green stalk anchors the plant to the rock, but this may not always be visible.

Colour   A lovely translucent green, often more emerald than the jewel. White patches on the older fronds show where spores have been released.

Texture   Smooth and slippery, but not unpleasant to touch.

Similars   The only confusion likely to arise between sea lettuce and other species is with *Enteromorpha* (see p. 24) and *Monostroma grevillei*, a similar translucent green but basically funnel-shaped with a split down one side. For the experts and those with sensitive fingers, *Ulva*'s frond is two cells thick whereas *Monostroma*'s is one cell thick (you can actually feel the difference).

Habitat   Not found on the upper levels of the beach but is otherwise common on rocky shores, especially where fresh water seepage occurs, or near sewage outfalls. In rock pools it is usually small sized and hidden beneath the larger wracks. Sometimes small plants grow on other large algae. It can tolerate moderate exposure to wave action; it is tougher than it looks.

Distribution   All over the world.

Abundance   Very common.

Picking time   In the spring if you particularly want to catch the vitamin C peak, but any time when the plants look in good heart for general eating.

Food rating   Very good. Well up on iron, vitamins A, B and C, and protein, plus all the other trace elements.

Uses As a vegetable, the large blade size making it easy to gather enough for a good meal in a short time. It is eaten raw, cooked or dried for future use, but the favourite way must be as a salad in its own right.

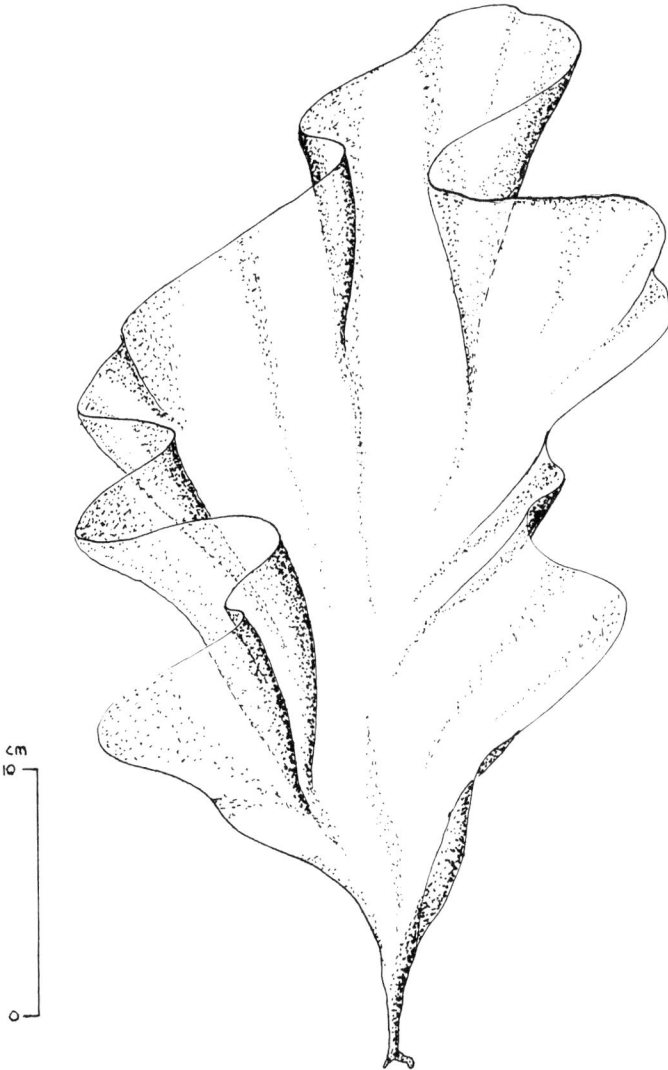

Bryopsis plumosa C.A. Agardh
Hen pen

Frond   Feather-like branches can reach 12 cm (5 inches) long, the side branchlets tapering towards the top of the plant and out to their ends. The branches are in one plane, arranged in two vertical rows on either side of the 'main stem'.

Holdfast   A creeping extension of the frond.

Colour   A glossy green, with a difference in shade between the sexes, female plants being dark green, the male plants more yellow.

Texture   Smooth, slightly hard to the touch, but the whole plant is limp.

Similars   In a class of its own among the greens. There are some browns which also have a feather pattern but you should be able to distinguish them quite easily.

Habitat   In the middle to low shore rock pools it often hides under other weeds, or hangs down from rocky overhangs; a 'shy' plant that's difficult to find.

Distribution   From the Mediterranean to the Baltic and up into the North Sea.

Abundance   Widely distributed and common, even though it is difficult to find initially.

Picking time   Hides in unlikely spots, but is well worth hunting down in the spring, when both vitamin content and taste are at their best.

Food rating   Flavour is difficult to categorize. To some it is a gourmet's delight, others dislike it. Its vitamin A and C levels are good, iron is moderate, as is protein content.

Uses   Small size is a problem, so *Bryopsis* tends to be used as a condiment.

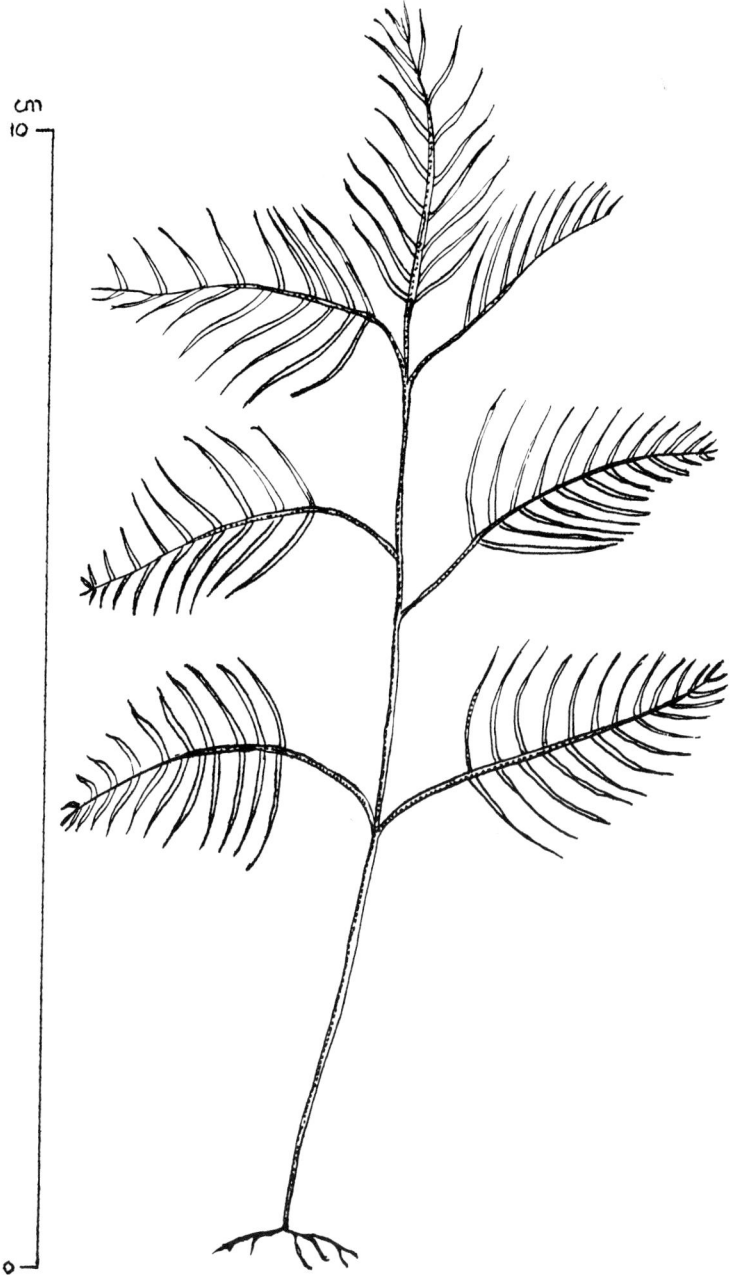

PHAEOPHYCEAE (Brown seaweeds)

Alaria esculenta (Linnaeus) Greville
Dabberlocks, wing kelp, henware, murlins

Frond A long leaf shape, up to 7 metres (23 feet) long with a definite midrib. The blade is narrow, 15 to 30 cm (6 to 12 inches) wide and is often torn and tattered. At the base of the frond lobe-shaped reproductive 'leaves' known as sporophylls sprout from the stalk.

Holdfast Looks very 'root'-like, branching.

Colour Yellow/brown/greenish, but never as green as the green weeds.

Texture The frond is soft, and feels delicate. Midrib is plump, rather fleshy and very flexible.

Similars You need to compare it with the large *Laminaria* to become familiar with the differences. *Laminaria* do not have the midrib.

Habitat *Alaria* has to be always submerged, even at extreme low spring tides, and likes very cold water. For such an apparently delicate plant, it is found on very exposed shores, anchored firmly to the rocks.

Distribution Cold water areas of the North Atlantic and the North Sea, especially round the Scottish coast.

Abundance Common where it occurs. In exposed conditions it replaces *Laminaria digitata* which prefers a somewhat quieter life.

Picking time Assuming you are well armed with snorkel, mask and wet suit (or a boat and hook) go for the highest vitamin C in spring. Other vitamins are at their best in other seasons. Cut the blade, leaving the stalk behind so that the perennial plant can regenerate. The plant reaches its greatest length in winter, but its greatest bulk in summer.

Food rating   Good. High in vitamins A, K, C and B, depending
on season. Iodine, bromine and trace element content also very
good. The diagram on p. 97 will give you some idea of when the
various peaks for vitamin content occur.

Uses   The midrib is sweet and crunchy, sporophylls are nutty,
but the frond frills have to be processed to remove an acrid taste.
Drying and smoking the frond turns it into a lovely food.

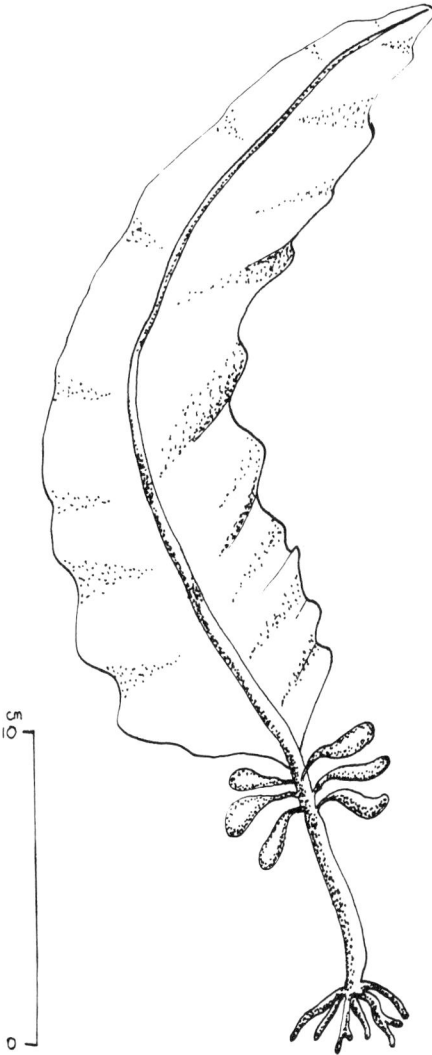

Ascophyllum nodosum (Linnaeus) Le Jolis. Knotted wrack, mussel wrack

Frond   An uneven, untidy looking plant. The frond starts off rounded, like a stalk, then gradually flattens towards its top, the main stem having uneven edges. There are many side branches, all of which have egg-shaped bladders in the body of the branch, and fruiting bodies at the branch ends in spring and summer. There is no obvious midrib.

Holdfast   Small fat lumps close together, with a stalk growing from each lump.

Colour   Generally olive green, but the fruiting bodies vary with sex, being olive in the female, and yellow in the male.

Texture   Tough but flexible, almost rubbery. The surface feels smooth except where eelworm (nematode) parasites cause small raised nodules.

Similars   Nothing close. The other wracks are distinct once you get to know them.

Habitat   Upper to middle shore in the wrack zonation. On gently shelving beaches and in estuaries. It does not like a site to be too exposed, but may be found on exposed shores in sheltered spots.

Distribution   Atlantic coast, English Channel and the North Sea.

Abundance   Given the right conditions of shelter the plants may blanket a whole beach.

Picking time   The sexes are separate, and each plant may live for many years, so cut carefully well above the initial stalk area, and before the fruiting bodies appear in the spring. If you would like to try and age a plant, rule of thumb measurement is one bladder for each year plus two years' initial growth.

Food rating   Good. High fat and oil content, helps it to resist drying out. Also of benefit to the consumer. Iodine is present but not overpowering.

Uses Delicate flavour after preparation such as steaming. Its great virtue is that it does not lose much vitamin A when stored correctly in the dry state. This, together with its abundance, makes it a favourite for animal feedstuffs and agricultural uses.

Note You may find red tufts of *Polysiphonia* growing on the *Ascophyllum* branches. All is edible, but make sure there are no hard lime or chalk bits in the holdfast area.

Chorda filum (Linnaeus) Stackhouse
Mermaid's hair, bootlace weed, fishing line

Frond Whip-like or string-like unbranched round frond, 8 metres (26 feet) long but only 6mm in diameter. The strand is hollow and gas-filled in an adult plant. Underwater, there appears to be a gelatinous cover which stands out from the plant along its full length. This may be a growth of fine threads of *Litosiphon pusillus*, another seaweed.

Holdfast A very small disc.

Colour Varying shades of brown from very dark to paler greenish brown.

Texture Slimy to the touch, but very flexible and very strong. If you bite it the frond feels crisp, like a good celery.

Similars Nothing quite like it, but a shallow water plant, *Chordaria flagelliformis* is often substituted for *Chorda filum* in recipes. It is smaller and has branches.

Habitat In permanent water, though shallow, down to about 20 metres (66 feet). It seems to prefer gravel or shingle bottoms, and plaice like to get among the strands. Such areas of gravel are usually quiet water zones.

Distribution From the Atlantic coast to the Baltic, in the quieter areas.

Abundance The description given of *Chorda filum* refers to the sporophyte generation, easy to identify and edible. It is abundant in favoured areas.

Picking time An annual plant, reaching its greatest length in high summer, which is the best time to pick it. Best collected in the deeper water, usually by snorkelling, but be careful not to get tangled. Weed washed up on shore is also used, but make sure it is clean.

Food rating Moderate. Good for starch, sugar and trace elements, but only moderate on vitamins. Iodine, boron and bromine are high, as is manganese.

Uses Fresh is best for food use. When dried it used to be a substitute for string.

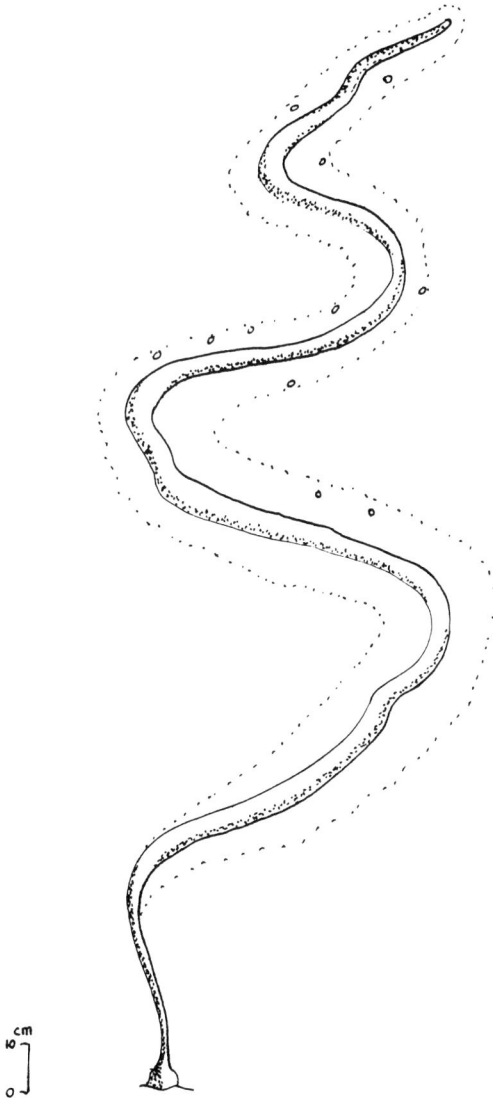

Ectocarpus species
Maiden's hair

Ectocarpus siliculosus (Dillwyn) Lyngbye is the species illustrated as an example in the various field guides.

Frond   Very fine, branched filaments up to 40 cm (16 inches) long. Strands are tangled together and form is very variable. It is a scruffy-looking plant.

Holdfast   A series of fine filaments that creep across the surfaces of rocks, other weeds, shells and stones.

Colour   Shades of brown from yellowish to greenish.

Texture   Firm, even crispy at times, but flexible. In summer it sometimes feels a little slimy.

Similars   There are many species of *Ectocarpus*, and identification by the layman is virtually impossible, as well as being unnecessary for food usage.

Habitat   Grows on rocks, stones and other seaweeds from the middle shore down to about 20 metres (65 feet), where wave action is not too fierce. Can tolerate a small variation in salinity, and likes its summer temperatures to be moderately warm.

Distribution   From the Mediterranean to the Baltic.

Abundance   Very common where it occurs, but needs hunting for.

Picking time   Spring to autumn, whenever it is found.

Food rating   High in iodine, and good quantities of iron, plus other trace elements. It has a sharp edge to its flavour.

Uses   Dried, to improve its flavour, then used as a condiment.

Scytosiphon lomentaria (Lyngbye) Link
Sugara, beanweed

Frond   A 10–30cm (4–12 inches) long, hollow, unbranched tube that tapers towards its tip. Along its length the tube walls pull in at intervals, so giving the effect of a string of sausages or beans. The frond is gas-filled and floats.

Holdfast   Microscopically small.

Colour   Greenish yellow brown. Colour varies with degree of exposure to sunlight.

Texture   Munchy but soft.

Similars   In a class of its own. Nearest look-alike might be a poor specimen of *Enteromorpha intestinalis*, but its colour would be obviously green.

Habitat   On rocks, stones, shells, other seaweeds, in shallow water and low on the shore. Can be found in deep rock pools where the temperature does not rise too high. Quite happy on exposed coasts.

Distribution   From the Mediterranean to the Baltic.

Abundance   Extremely common. The plant has a peculiar survival pattern. In the summer, when the temperature rises, specimens are found higher up the shore than in winter. This may perhaps be connected with reproduction.

Picking time   Most plentiful in the winter except towards the edges of its range in the Baltic.

Food rating   Very good. It has plenty of sugar and starch, a high iron content plus all the usual trace elements, vitamins and minerals. Flavour is very good, a pleasant bean-like taste that is improved by drying.

Uses   Japanese love it in soup. Use fresh or dried.

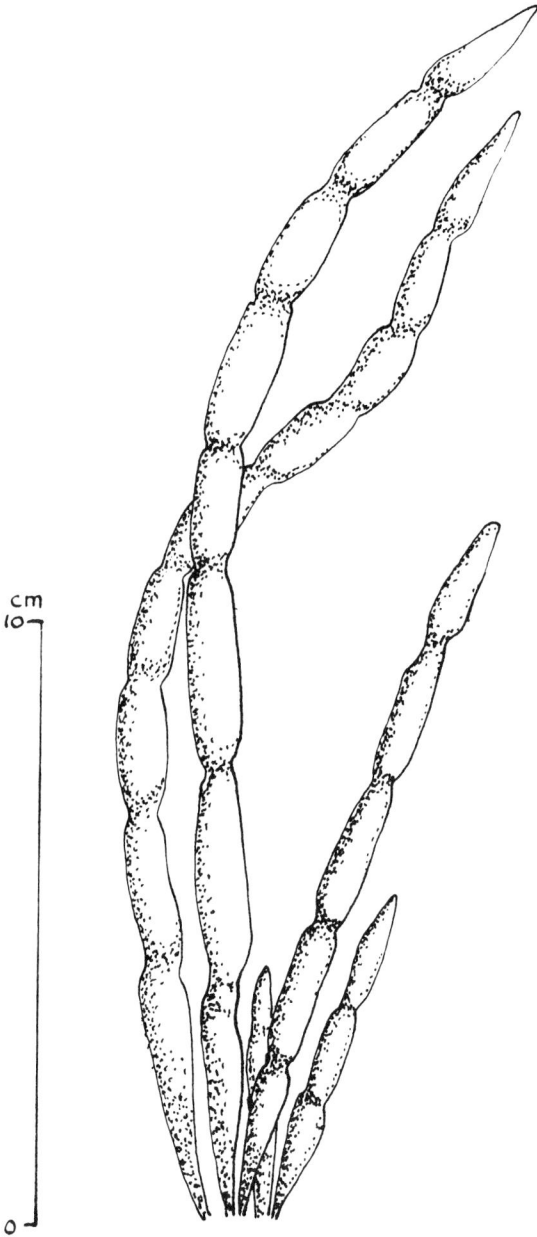

Fucus vesiculosus Linnaeus
Bladderwrack, popweed, pig weed, rock weed, red or dyer's wrack

Frond   A tough weed that has branching, flat, wavy-edged offshoots (not toothed) and a conspicuous midrib. Each branch has two or sometimes three gas bladders grouped together in the body of the frond, and in due season, reproductive bodies at the branch tips.

Holdfast   A blob or lumpy disc. Cells on the underside of the holdfast secrete a kind of glue with which to stick the plant to a rock.

Colour   Brown and olive-green. Reproductive bodies can be yellowish in the male plants, more green in female plants.

Texture   Tough and leathery.

Similars   Comparison with *Ascophyllum nodosum* will soon dispel any confusion. A Mediterranean species, *F. virsoides*, does not have air bladders.

Habitat   A middle shore species in the zonation scheme of wracks. It does not like very exposed sites, preferring rocks where the waves have lost much of their power before reaching the weeds.

Distribution   From the Atlantic coast to the Baltic. A plant of cooler waters.

Abundance   Very widely distributed and very common.

Picking time   A perennial plant, so cut carefully. The best vitamin A level is in the summer, but vitamin C is highest in the autumn.

Food rating   Good. Very good levels of protein and vitamin A. High in phosphorus, bromine, magnesium and iodine. Good on sugar, starch, fats and zinc. It has a sweet and lovely flavour.

Uses The plants themselves are not eaten, instead all the goodies are stewed out and used as a tea, or a broth. Weeds are usually dried for storage, and lose very little of the nutrient content as a result of this treatment.

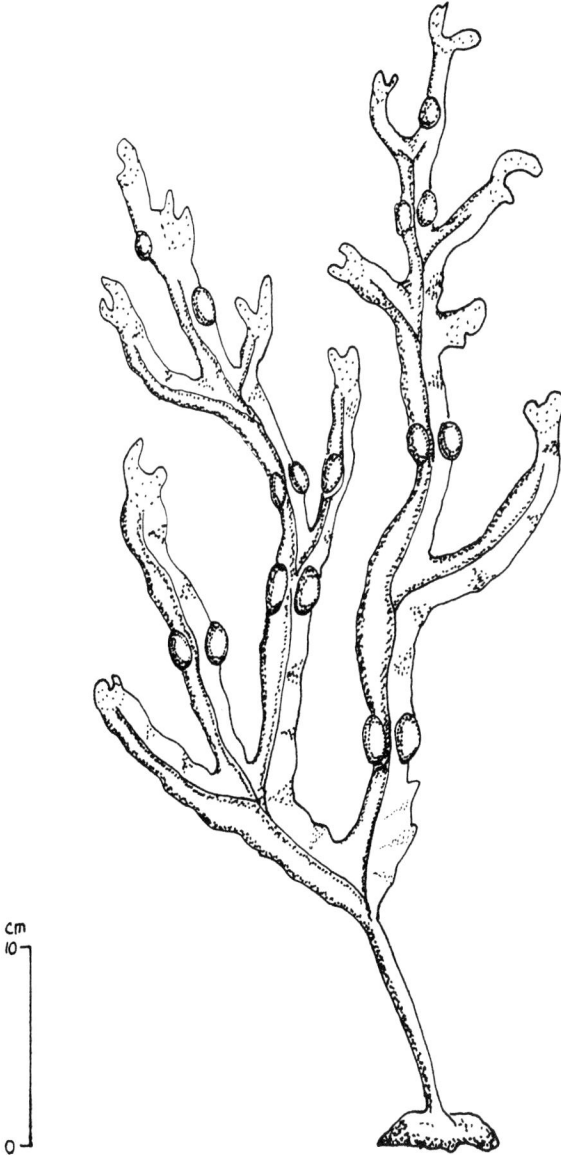

Fucus serratus Linnaeus
Toothed wrack, chicken balm, bitter wrack

Frond A tough, irregularly branching flat frond on a short stalk. There is a thick midrib, and the edges of the frond are toothed. Over the surface of a mature plant, tiny tufts of 'hair' give it a spotty appearance. No air bladders. Whole plant can grow to more than 60 cm (24 inches) long.

Holdfast Strong and root-like.

Colour Colour difference between the sexes is not very marked. Male plant is orange-brown, female more brown.

Texture Tough, smooth, strong plant that can feel very slippery.

Similars Most of the other wracks are similar, but a quick comparison of fresh plants on the beach will soon clear up any confusion. *Fucus spiralis* is the closest in form, and is used the same way.

Habitat Has a distinct zone in the wrack shore profile (see Chapter 6), covering the lower to middle area where it is not exposed to air for any great length of time. Many other plants and animals grow on the fronds of *Fucus serratus*.

Distribution From the Atlantic coast to the Baltic on all but the most exposed areas.

Abundance Very common.

Picking time The plant fruits in winter, so pick from the spring onwards, cutting above the stalk.

Food rating Poor. This is not a food wrack for human consumption. Very high in iodine, high in carbohydrate, with moderate protein and trace element content. The iodine makes it bitter as food, but nevertheless it is a useful weed.

Uses Dried in bulk, then used in medicine, and as a valuable animal food or manure plant.

Laminaria digitata (Hudson) Lamouroux
Kelp, tangle

Frond   A wide blade, divided into flat, strap-like 'fingers', rising from a long flexible stalk. The stalk is oval in section and thick. Plants can grow to a metre (3 feet) or more in length.

Holdfast   Root-like, with very many branches that spread out across the rock surface.

Colour   Rich brown, rarely yellowish. Very shiny.

Texture   Smooth to touch, very pliable and slippery.

Similars   If you find a weed with a rough stalk, then it is *Laminaria hyperborea*. A stalk with a wavy edge and bulbous outgrowths from the holdfast is *Saccorhiza polyschides*. Both are deeper water species than *L. digitata* and most likely to be found cast up on the beach after storms, rather than growing.

Habitat   In cold, deep water down to 5 metres (16 feet) or more, but will grow up to a level where the tops may be exposed on a very low spring tide if other conditions are right.

Distribution   From the Atlantic to the Baltic.

Abundance   Very common.

Picking time   Around the European coast it is a perennial. New growth appears at the base of the frond in early spring, and the older upper parts are eventually shed. American plants are smaller and seem to be annual growth on perennial holdfasts. Cut frond only, leaving some attached to the stalks. Cast weed is acceptable provided it is fresh and uncontaminated.

Food rating   Good. A pleasant tasting plant that contains a natural version of sodium glutamate, so stimulates the taste buds. Very high in iron, iodine and zinc, good quantities of all the other minerals, vitamins and trace elements.

Uses   Sun dried or smoke dried, *Laminaria* is a tasty and beneficial food.

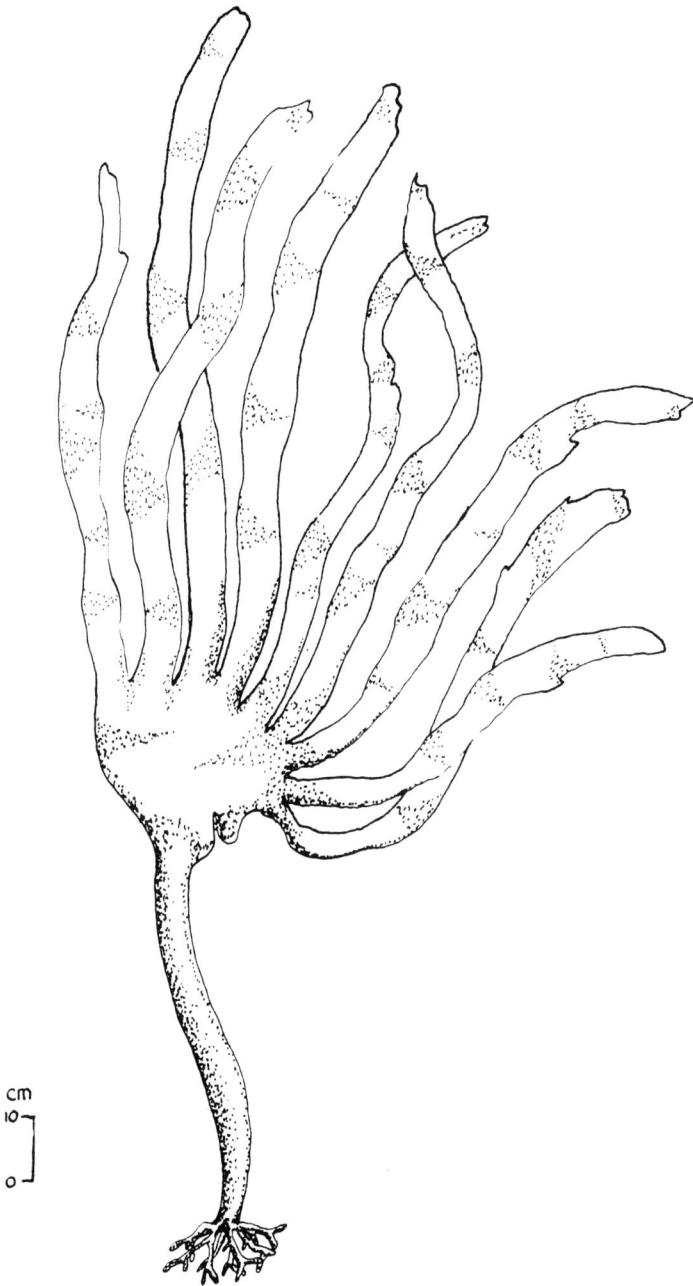

cm
10
0

Laminaria longicruris
Oarweed, another 'kelp'

Frond   Long, broad, single unbranched, undivided blade up to 3 metres (10 feet) long. No ruffles or wavy edges, no midrib or central crumpled effect on the blade.

Holdfast   Large, strong and root-like.

Colour   Olive brown.

Texture   Smooth, pliable and slippery.

Similars   There are many oarweeds, all looking almost identical to each other. In addition, broken-off fingers of *L. digitata* may be mistaken for oarweed. From a food point of view, this does not make any difference. Watch out for *Laminaria saccharina*, the sugar weed that has a special taste all of its own.

Habitat   A deep-water species that occasionally ventures to near the low spring tide mark. Often cast up on the beach.

Abundance   Common.

Picking time   Summer is the best, especially if you can cut fresh weed while out snorkelling. When cutting, leave the stalk with a piece of frond attached.

Food rating   Good. Another kelp, so all that has been said for *Laminaria digitata* applies to this weed also.

Uses   As a food, dried or smoked over oak chippings.

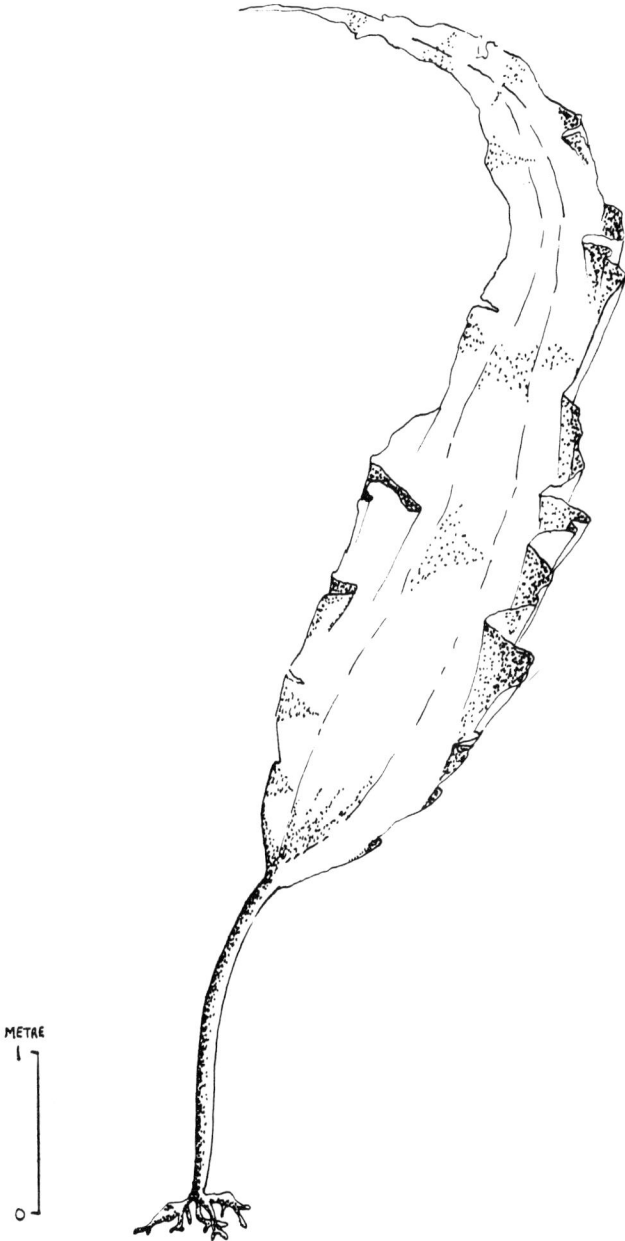

METRE

Laminaria saccharina Lamouroux
Sugar kelp, weather weed, sea belt

Frond   A crumpled-looking, long, unbranched blade with no midrib, growing to 3 metres (10 feet) long. The edges of the frond are wavy. Compared with other *Laminaria* the stalk is thin. It is smooth and round and can be up to a quarter of the frond length.

Holdfast   Root-like and branching on several levels from the stalk, giving a layered effect.

Colour   Varies from yellow-brown to rich chestnut.

Texture   Thick, smooth and pliable.

Similars   In its very young stage, *L. saccharina* can be mistaken for *Petalonia fasciata*, although colour is the deciding factor in identification. Other oarweeds do not have the crumpled look.

Habitat   Below low spring tide level down to 20 metres (66 feet) or more in areas where there is some shelter from winter storms.

Distribution   From the Atlantic to the North Sea.

Abundance   Common.

Picking time   Although it is a perennial plant, it does not regenerate easily after cutting. Left to its own devices it will renew itself by growing from the top of the stalk. So if you cut fronds, leave a good piece still attached to the stipe. Otherwise, gather the weeds cast up by storms. Summer time is best, for that is when the sugar, from which it gets its name, is at a maximum.

Food rating   Good. Besides all the usual benefits of kelp (see Chapter 3), *L. saccharina* has a natural sugar, mannitol. This crystallizes out on the plant's surface when it is dried.

Uses   Used dry or fresh for food; also ideal for making alcohol.

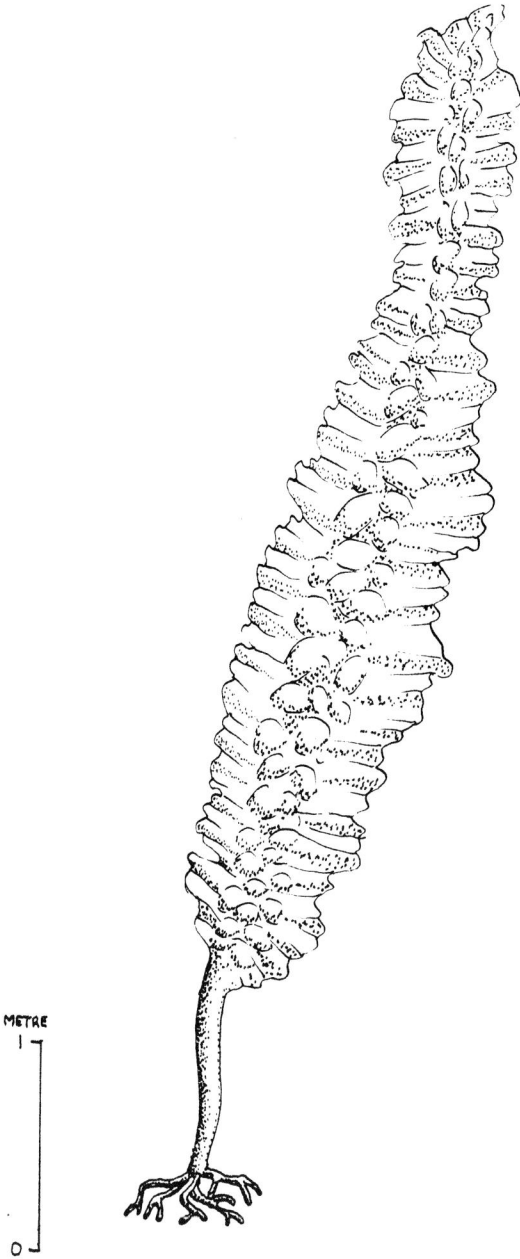

METRE
1
0

Pelvetia canaliculata (Linneaus) Decaisne and Thuret
Channelled wrack, cow tang

Frond   A branching plant that appears to be made of an open-sided tube. The groove or channel is on one side of the frond only. There is no midrib and no air bladders. At the tips of the branches, granular reproductive bodies swell out, looking like tiny cones. The whole plant is small, only 15cm (6 inches) long in good conditions.

Holdfast   A blob made of tightly felted strands.

Colour   Green brown, turning black when dried.

Texture   A coarse plant, slightly rough to the touch, but very strong.

Similars   Position will distinguish it from the other wracks, and the channel is a distinctive feature.

Habitat   Appears to be almost a land plant in some areas, extending up into the splash zone above the spring tide level. Generally it is found on the upper shore, hanging down from rocks so that the channel side is against the rock face. At neap tide times, the plants may be out of the water for days.

Distribution   From the Atlantic coast to the North Sea.

Abundance   Widespread and common, often very abundant when conditions are just right.

Picking time   Spring is best. It has the advantage of being available whatever the set of the tide.

Food rating   Moderate. It has a very strong flavour, which some enjoy while others find overpowering, and a high fat content.

Uses   Sun drying improves the flavour, then it makes an excellent condiment.

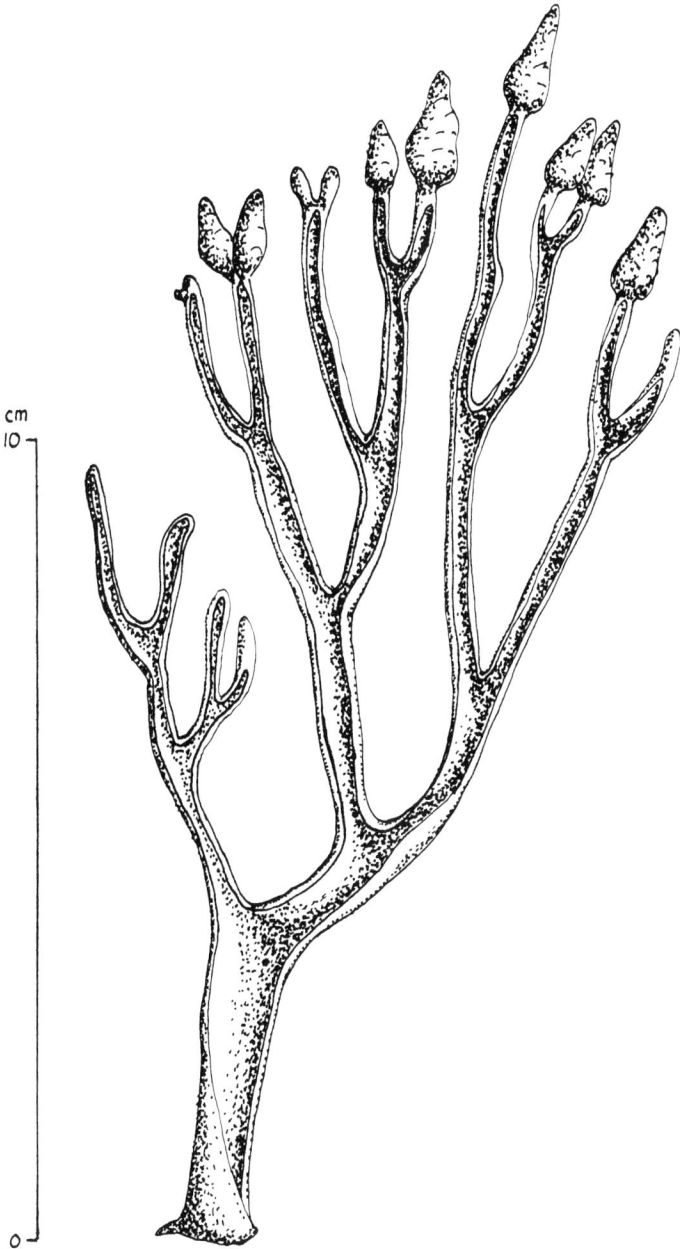

Petalonia fasciata (O.F. Müller) Kuntze
Petal weed, patty

Frond   Very short, thin stalk that widens out into a simple blade, up to 30 cm (28 inches) long by about 6 cm (2½ inches) wide. The edges of the frond may sometimes be a little frilly. No spots or other marks.

Holdfast   Small disc.

Colour   Greenish brown.

Texture   Smooth and glossy, yet tough.

Similars   There are other similar species. A slightly larger plant is *Punctaria latifolia*, which is spotty. If in doubt, the glossy frond is a good indication of *Petalonia fasciata*. May also be confused with very young *Laminaria* plants, but here the tiny stalk of *Petalonia* is the clue.

Habitat   From the mid shore down to low tide level, in pools and sometimes buried in the sand.

Distribution   From the Mediterranean to the Baltic.

Abundance   Common where it occurs, but not well documented.

Picking time   A winter species, very unusual. Starts growing in November, reaches a maximum in mid-winter, then gone by April. It is an annual plant.

Food rating   Moderate. It has good levels of sugar and starch, and its flavour is pleasant.

Uses   Fresh or dried, it is used as a substitute for *Porphyra* or *Undaria*. As a winter weed, it was once used to provide fresh vegetables during the 'hungry gap' at the end of the cold weather.

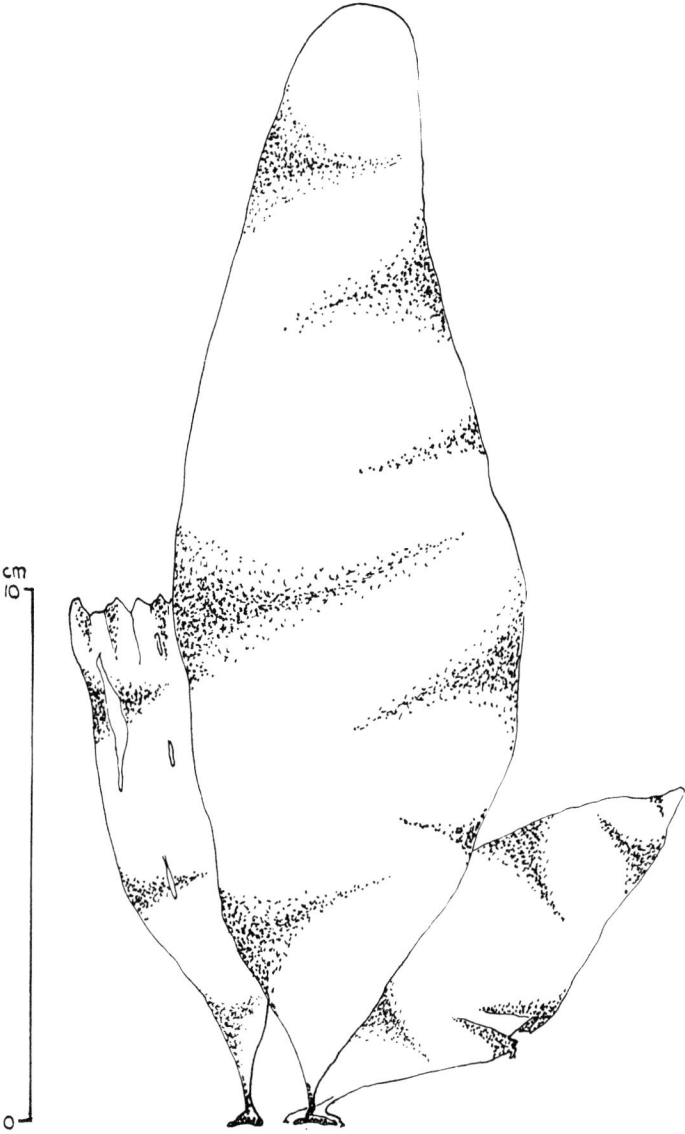

Desmarestia aculeata Lamouroux
Desmarestia ligulata (Lightfoot) Lamouroux
Desmarestia viridis (O.F. Muller) Lamouroux

Sea sorrel. The three species are shown in one diagram.

Frond   Opposite or alternate branching from a central stem, giving the plants a more or less feathered appearance. The smallest is *D. viridis*, only 30 cm (12 inches) long, whereas the other two can reach 180 cm (71 inches).

Holdfast   Small neat disc.

Colour   Generally olive-brown, changing to green as the weed decomposes. Older plants of *D. viridis* sometimes look reddish.

Texture   Soft young plants become more rigid as they age. *D. aculeata* has a thorny appearance.

Similars   Many feathery browns and reds, but *Desmarestia* gives itself away by bleaching out any weeds near it.

Habitat   On rocks, from the lower shore down to shallow water, and sometimes in pools.

Abundance   Fairly common.

Picking time   Not usually picked. Contains esters of sulphuric acid which can cause severe stomach upsets.

Food rating   Not recommended.

Uses   As a source of acid in dilute form for small domestic industry such as dying wool.

RHODOPHYCEAE (Red seaweeds)

Ahnfeltia plicata (Hudson) Fries
Landlady's wig

Frond   Stiff, many branched, with thin fronds forming a tuft-like plant. The tuft is usually about 10 cm (4 inches) in diameter and up to 15 cm (6 inches) tall.

Holdfast   A thin film of fronds that encrusts the rock to about a quarter of the diameter of the tuft.

Colour   Dark red to almost black.

Texture   Like fine wire, hence its common name of landlady's wig.

Similars   Nothing quite so wiry. The American species *A. gigartinoides* is a much larger plant, and more brownish red.

Habitat   From the middle shore down to deeper waters, in exposed areas, where it remains attached to rocks. Also found in pools, and floating in mats where the water is more quiet – this latter could just be a result of storm-torn weeds being washed into the calm zones. Weed can continue to grow even when removed from its holdfast.

Distribution   From the Atlantic to the North Sea.

Abundance   Widely distributed and locally common.

Picking time   A perennial plant that can be gathered from spring to autumn. Cast weed is often gathered but the best plants are in deeper water, and are gathered by snorkellers. The plant is very resistant to decay.

Food rating   Despite its offputting appearance on first acquaintance, the weed is an excellent agar source. It is much used in the U.S.S.R. but is not abundant enough around the British coast to be commercially viable for agar production. Contains starch, sugar and all the usual trace elements. Taste is pleasant, and the plant is munchy.

Uses   Directly as a food, raw or processed. As an agar source after suitable treatment.

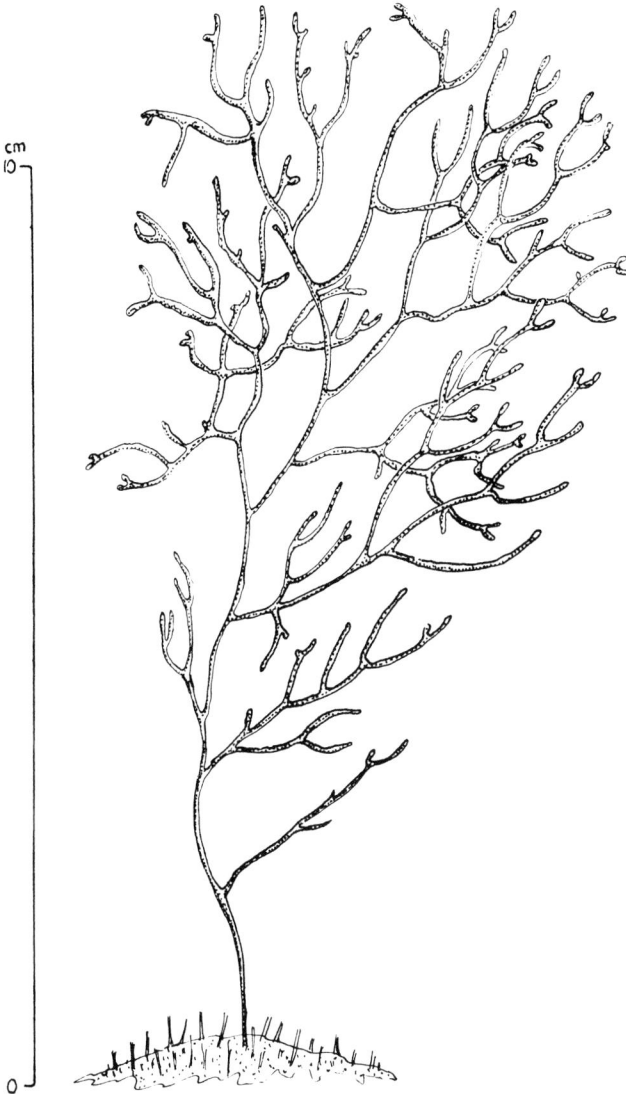

Bonemaisonia asparagoides (Woodward) C.A. Agardh
No common name

Frond   A long cylindrical or slightly flattened stem that bears alternate side branches which decrease in size towards the top of the plant. The branches bear sub-branches, which in turn carry branchlets. These branchlets are approximately the same size over the whole of the plant. Fronds can reach 25 cm (10 inches) long.

Holdfast   Small and disc-like.

Colour   Red, sometimes brilliantly so.

Texture   Smooth, not slimy. The plant is very floppy but not delicate.

Similars   Several other red branchy weeds, but none so precisely arranged.

Habitat   A shallow-water species that likes to be able to wave in the 'sea breeze' set up by current movement. Rarely in rock pools at the bottom of the shore.

Distribution   From the Mediterranean to the North Sea.

Abundance   Not well known but probably common. The seaweeds of the shallow waters are rarely covered by ordinary shore-bound identification books.

Picking time   I refer you to the section on reproduction at the beginning of this chapter, in particular to the part about one sexual phase of a plant being so different in shape to the other asexual phase that they are identified as two separate species. This has happened to *B. asparagoides*, which is now known to be the sexual reproductive phase of *Hymenoclonium serpens* (Crouan frat) Batters. Try saying that mouthful after a night on the strong ale. *H. serpens* is not illustrated because it is extremely unlikely you will ever find it. Pick *B. asparagoides* sparingly and in the late spring. You may need to be a scuba diver to get this one.

Food rating   Good. All red weeds are reasonable food plants. This one is strong on trace elements, but has iodine and other esters in it which give a strange flavour to the raw weed.

Uses   Drying improves the flavour. Use as a condiment, or rehydrate and use as a vegetable.

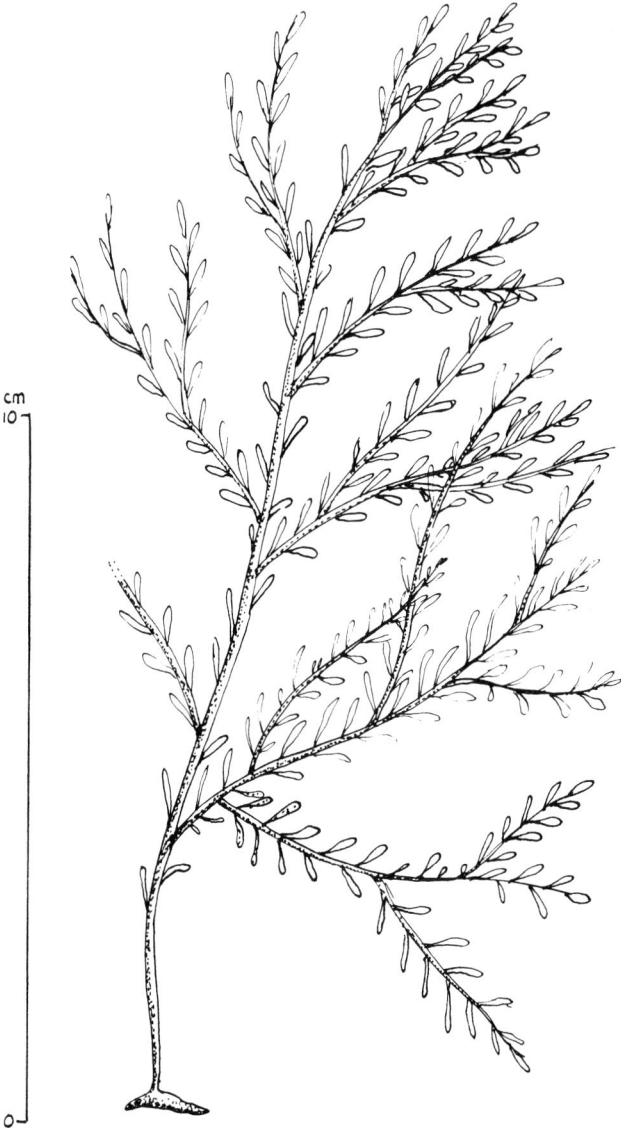

cm
10

0

Chondrus crispus Stackhouse
Irish moss, carragheen, jelly moss, sea moss, bejin gwenn, pioca and many other regional names

Frond   Flat, wide, with no midrib or channel. Edges are not curled or rolled inwards. Width of frond varies from broad to narrow but never thin enough to be considered tufty or wiry. Each branch divides into two, then two again and so on ('dichotomous branching'). The whole plant is rarely more than 15 cm (6 inches) tall, with a curly mop-head appearance. Fruiting bodies appear in winter as little lumps on the frond.

Holdfast   Small button or disc.

Colour   Red to purple, sometimes so dark as to appear purple-brown or black. In brightly lit pools the frond will go green, but the stalk will remain dark coloured. Occasionally the plant will be bleached out to a creamy white. Sometimes, a soft plant in a quiet pool will appear to be a lovely iridescent purple, but this vanishes when the frond is taken from the water.

Texture   Soft but crunchy, generally referred to as 'cartilagenous'. Stiffer plants are found in the more exposed areas, softer ones in pools.

Similars   The plant most often confused with Irish moss is *Gigartina stellata*, but from a food point of view there is no problem, because one is often used as a substitute for the other.

Habitat   Along the lower shore and into shallow water, occasionally in pools which do not dry out. Settles on rocks, shells, pier piles, old wrecks.

Distribution   From the Atlantic to the North Sea.

Abundance   Very common and usually plentiful. The broader-bladed, softer plants are found where water is quiet and the beach sheltered.

Picking time   Perennial plant; can be picked all the year round, but the best time is in spring and summer when the vitamin A content is at its highest.

Food rating   Very good. Rich in trace elements and vitamins, plus protein, fat and carbohydrate, in fact almost a complete food.

Uses   Multitude of uses both as a food and as a gelling agent. Bleached and dried before use, and available in most wholefood shops.

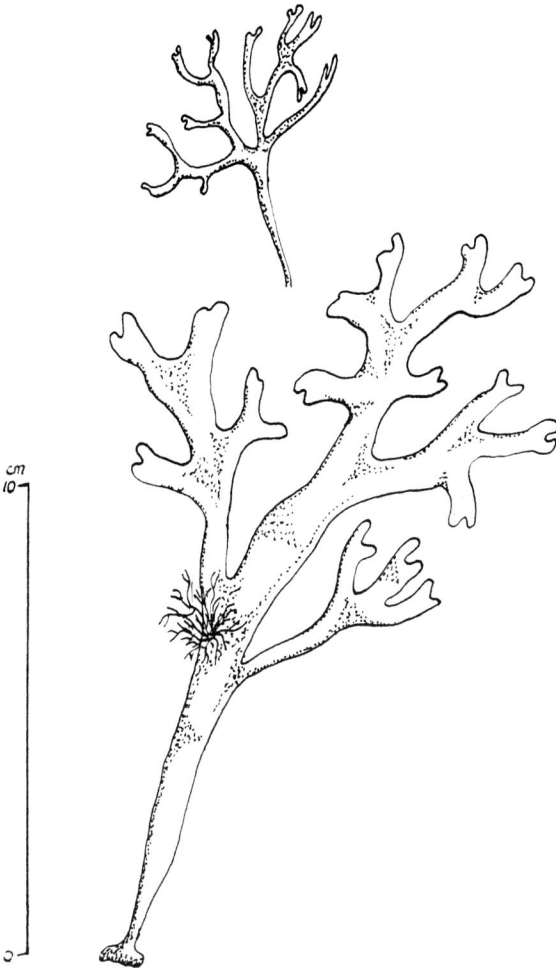

Gelidium latifolium (Greville) Bornet and Thuret
Gelidium sesquipedale Thuret
Jelly plant

Frond A flat, ribbon-like main branch from which come thin branchlets. Whole frond is up to 10 cm (4 inches) long.

Holdfast Creeps over the rock face, many fronds arising from it, to give a 'turf' effect.

Colour Rich deep red, crimson, sometimes purple.

Texture Cartilagenous, soft yet crunchy.

Similars The overall pattern is the prime identification point. In south-west England, *G. sesquipedale* is dominant, *G. amansii* in America. The warmer waters of south-west England down to the Mediterranean nurture *halarachnion* species. A scruffy-looking version of *G. latifolium* is likely to be *Calliblepharis ciliata*, characterized by its root-like holdfast. It has the same uses for food but is not such a good agar producer.

Habitat From the middle shore down into shallow water, on rocks, and under overhanging rocks in pools.

Distribution From the Atlantic coast to Ireland and the English Channel. Prefers warmer waters but occasionally turns up on the Scottish coast.

Abundance Widely distributed and very common in places.

Picking time A perennial plant, annual fronds arising from the holdfast. It has a complex life history which is not yet fully understood. *Gelidium* is cut any time when the plant is growing (i.e. when the annual frond is growing up from the perennial holdfast).

Food rating The fresh weed has a definite mousey flavour which is removed by processing. High in iodine, sugar, starch and trace elements.

Uses Chief sources of agar. Drying, bleaching and boiling are used to improve the flavour.

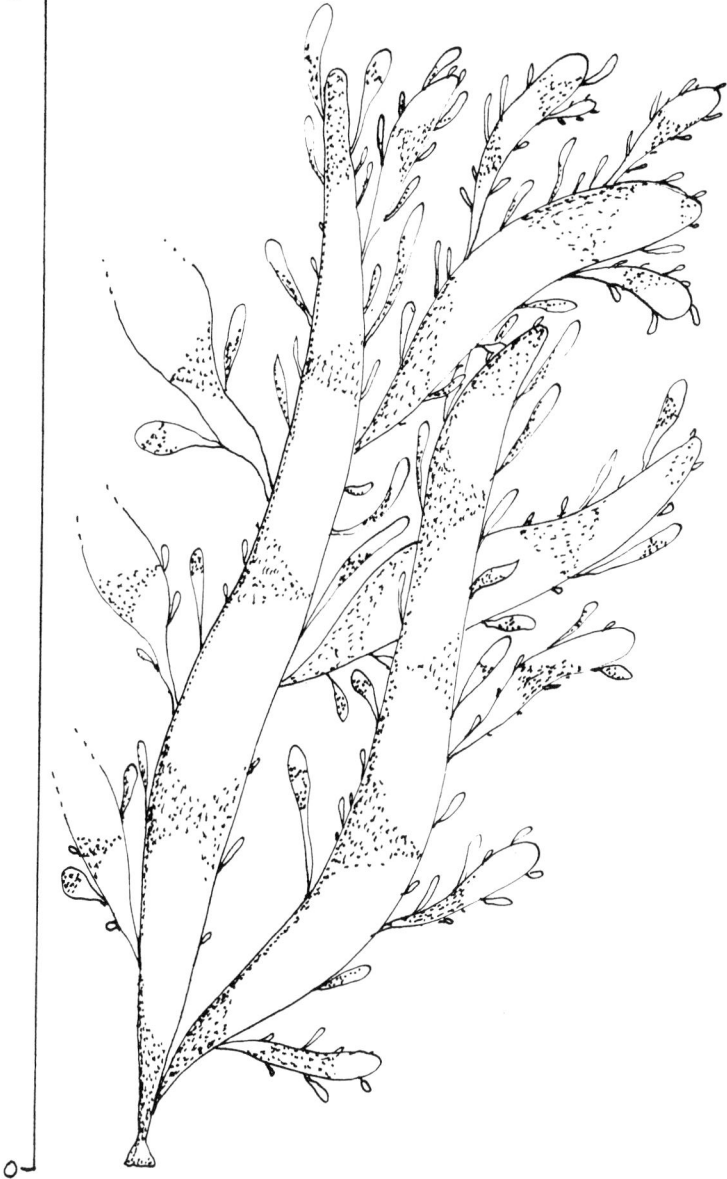

Gigartina stellata (Stackhouse) Batters
No common name

Frond   Flat blade with rolled edges, giving it a slightly channelled appearance. Branched, so that whole plant forms a tuft. Frond narrows down towards the holdfast. Fruiting bodies form as small lumps on older plants. Plant stands about 20 cm (8 inches) high.

Holdfast   A tiny disc.

Colour   Brownish purple/red.

Texture   Stiff, smooth surface except for knobs of fruiting bodies.

Similars   Often confused at a superficial glance with *Chondrus crispus*, but the confusion is not important for food use.

Habitat   On rocks at the lower end of the shore, where it may be so abundant that it forms a broad band.

Distribution   From the Atlantic to the North Sea, but most plentiful on the west coast.

Abundance   Very abundant and widely distributed.

Picking time   Mid spring is best, for the highest vitamin C content, otherwise whenever the plant is available.

Food rating   Very high in vitamin C, and has a nice flavour once processed.

Uses   A carragheenan source (carragheen or Irish moss gave its name to the jelly derived from it, but the jelly can also be made from other seaweeds). Used during the war when *Gelidium* agar was not available from Japan. Sun dried.

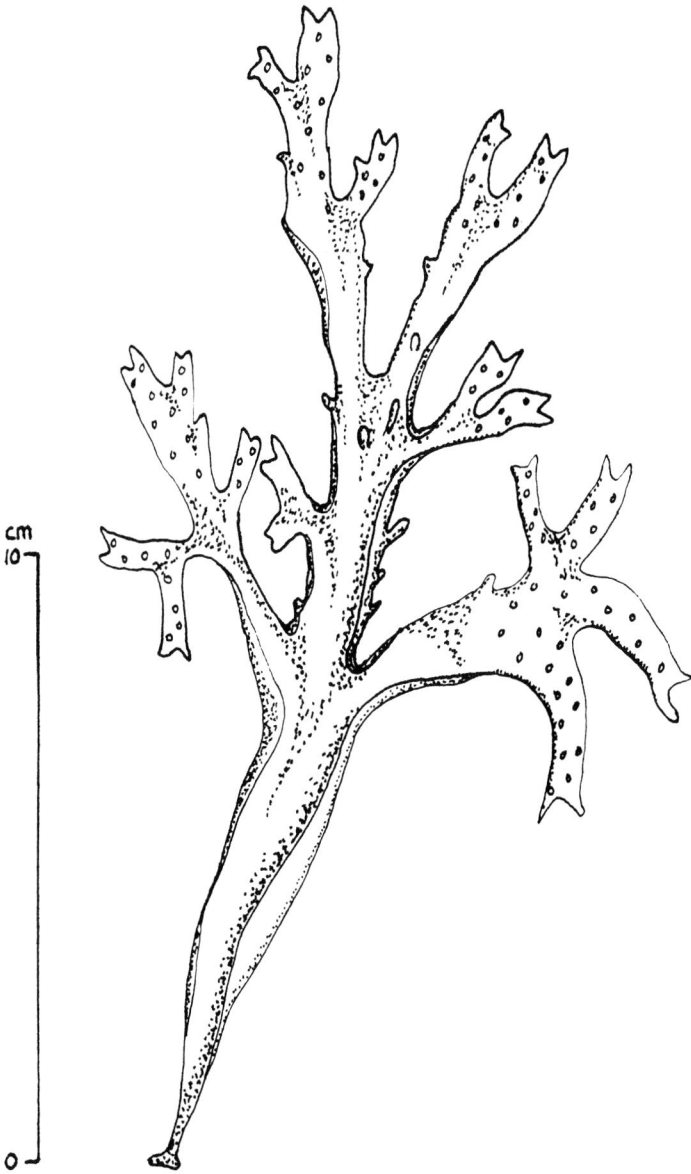

Gracillaria verrucosa (Hudson) Papenfuss
Old name in some books is *Gracillaria confervoides*
No common name

Frond A stringy stem with irregular branches from which arise slender branchlets that taper at each end to smooth slender points. Knobbly, rough, wart-like reproductive bodies dotted over the plant. Frond can reach 50 cm (20 inches) long.

Holdfast A tiny disc with many little projections, from which rise many fronds.

Colour Red.

Texture Soft and munchy, very soft when young.

Similars A similar unknobbly plant is *Dumontia incrassata*, more brownish in colour. There are several other bunchy weeds but identification should be fairly straightforward. In colder waters up into the Baltic, *Cystoclonium purpureum* has a wider distribution. It has a root-like holdfast and reproductive bodies on the tips of its branches, and can be substituted for *G. verrucosa*.

Habitat On the middle shore, it can be found growing up through the sand from buried rock, and in gravelly areas.

Distribution From the Mediterranean to the English Channel.

Abundance Widely distributed but locally common.

Picking time Perennial holdfast with annual frond that dies off in winter. Best picked in spring before the fruiting bodies start into activity. Cast plants can also be gathered.

Food rating Good. High in nutrient value, especially zinc. Very high quality agar.

Uses An agar producer. Eaten raw, sunbleached or blanched. When dried it looks like string.

66

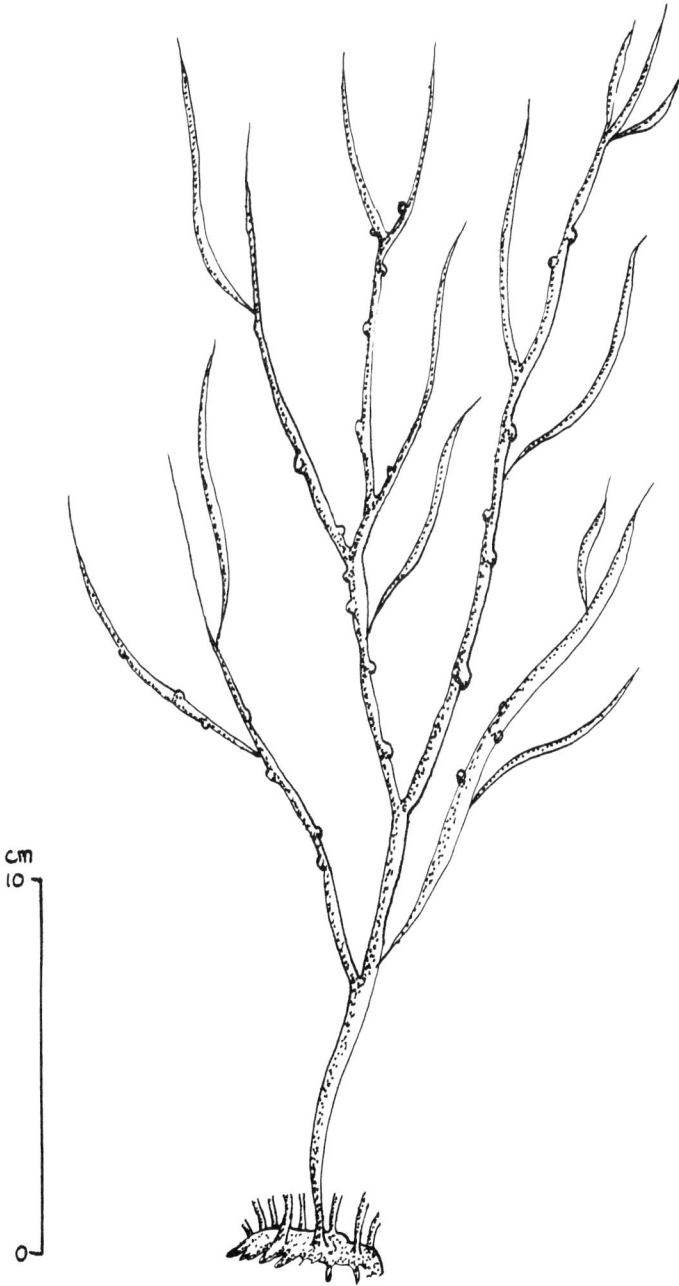

Nemalion helminthoides (Velley) Batters
Sea noodle, Turkish spaghetti (Italian name)

In old books the names may be *N. elminthoides, N. multifidium*

Frond Worm-like branching stems that taper to blunt points. Diameter of the stem is about 2 cm (¾ inch), and the plant can grow to 25 cm (10 inches) long. Variants may be less branched or unbranched, hence the reason it had several different names.

Holdfast A minute disc.

Colour Brown/red/purple.

Texture Cartilagenous, slippery and sometimes like jelly.

Similars In a world of its own.

Habitat Likes exposed positions on the middle shore, often in pools.

Distribution Atlantic coast. Rare in the North Sea.

Abundance Where it does occur it is very common, but the crop is variable.

Picking time An annual plant that reaches its peak growth in the summer. Pick any time.

Food rating Good. Starch and sugar give it a pleasant mild taste that is popular.

Uses Dried or fresh, as a vegetable.

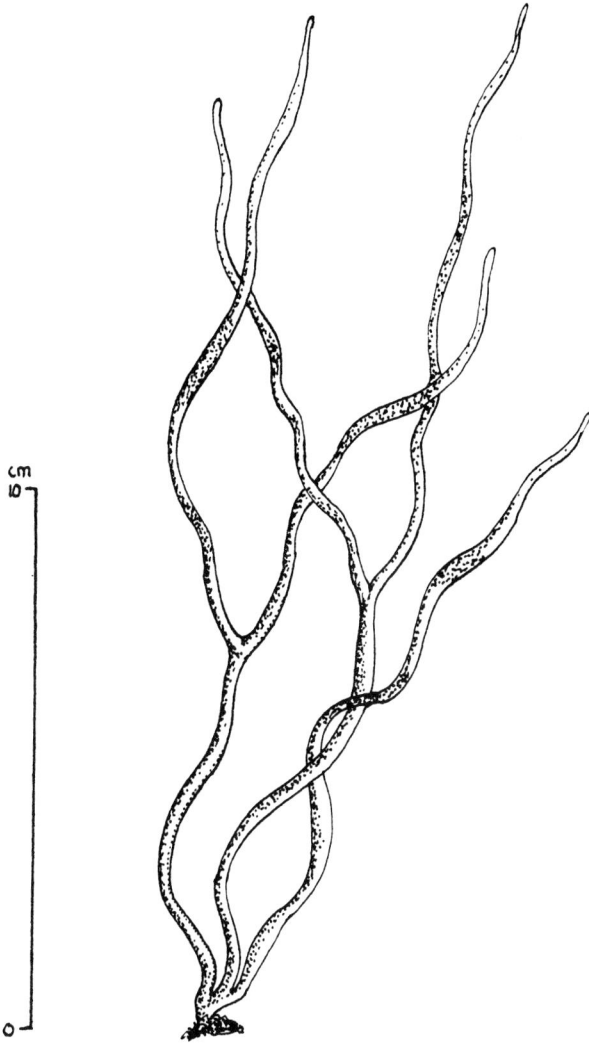

Rhodymenia palmata (Linnaeus) Greville
Dulse, dillisk, red kale

Referred to as *Palmaria palmata* in some publications.

Frond   A wide, fan-like blade that has many side blades and bladelets in older parts of the frond. There is no stalk. Plant may reach 30 cm (12 inches) long.

Holdfast   A wide, thin disc.

Colour   Deep red to purple.

Texture   The plant looks tough but in fact it is quite soft. Smooth surface that is slippery to the touch.

Similars   On the lower shore and in shallow water *R. palmata* is likely to be confused with *Dilsea carnosa*, until you look closely at the two plants side by side. *D. carnosa* has the same use as *R. palmata* and some consider it to be the finer plant. Identification point is the presence or absence of a stalk. *Calliblepharis cilliata* is also a confuser, but food use is the same.

Habitat   Grows on rocks and on the holdfasts and stalks of *Laminaria*, from the middle shore down to low spring tide level.

Distribution   From the Atlantic coast to the North Sea.

Abundance   Widely distributed and abundant.

Picking time   Local lore has it that you do not pick dulse when there is an R in the month. Perennial plant, gathered in summer and early autumn.

Food rating   Good. Very high in protein, fat and vitamin A. The highest concentration of vitamin A occurs in mid summer, highest vitamin C in early autumn. Rich in other food substances and trace elements. Rich nutty taste.

Uses   A carragheenan source. Commercially exploited. Made into food and alcohol just about everywhere.

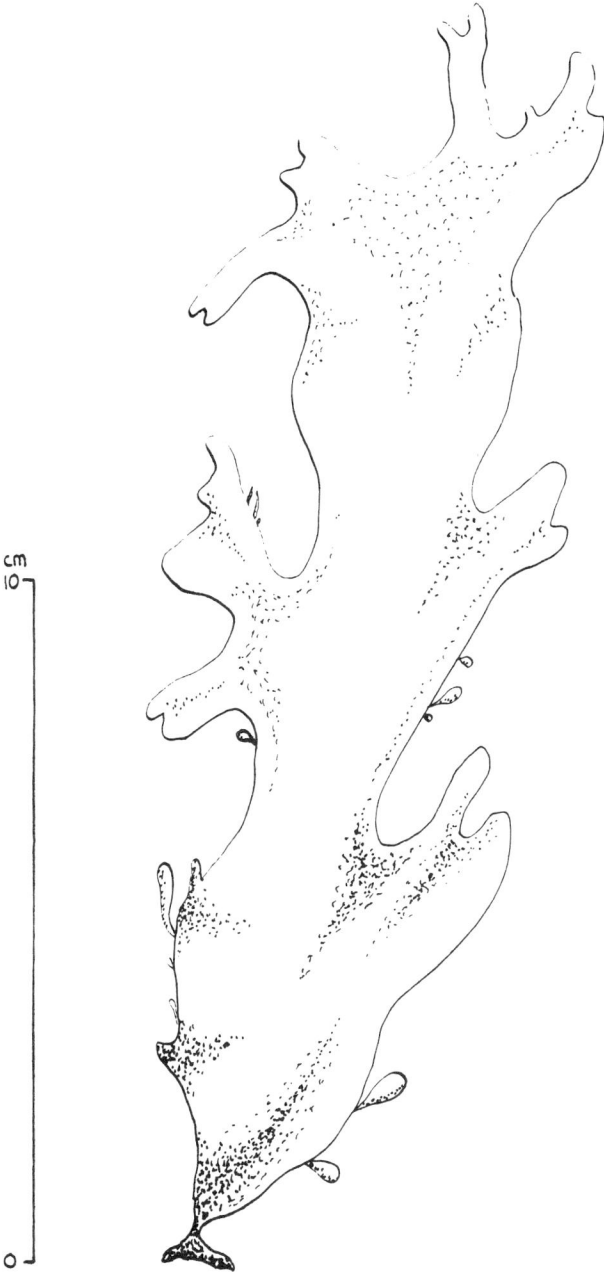

Porphyra umbilicalis (Linnaeus) J.G. Agardh
Laver

Frond   An irregular, thin, almost transparent frond that is attached at one point without a stalk. The blade is sometimes lobed, but generally considered to have no definite shape. Often damaged due to sea and sand movement. About 20 cm (8 inches) in diameter in a good frond.

Holdfast   Seems to grow straight out of the rock.

Colour   Red, purple, often green in the centre of the frond. Black when dry.

Texture   Smooth, gelatinous, floppy.

Similars   Green sea lettuce. The two plants are interchangeable as food. Also there are about five other species of laver, all used the same way.

Habitat   Exposed beaches covered with sand, not the sort of place you would expect to find such a delicate-looking plant. Grows on rocks and stones.

Distribution   From the Mediterranean to the North Sea.

Abundance   Widely distributed and very common.

Picking time   An annual plant. As for dulse, do not pick when there is an R in the month.

Food rating   Good. The smaller *Porphyrae* are thought to have the better flavour. High in protein. Very good vitamin B content. A good all-rounder for nutrients.

Uses   Eaten fresh or dried. Women in South Wales who regularly ate laver bread were found to have lower breast cancer rates than in the rest of Britain.

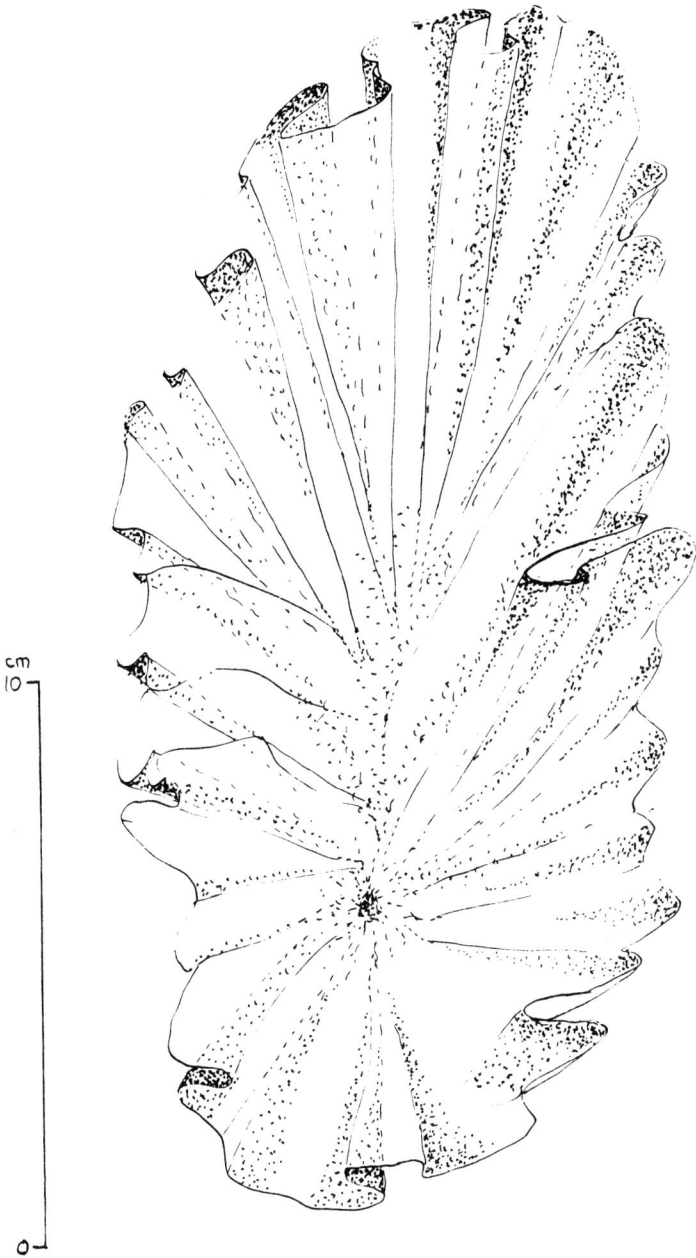

Polysiphonia lanosa (Linnaeus) Tandy (= *P. fastigiata*)

Frond   Branched rounded filaments up to 9 cm (3½ inches) in length. Many filaments arise from a common base to give the plant a bushy or tufted appearance.

Holdfast   Very fine rootlets which invade the host tissue.

Colour   Red to red-brown, often very dark red towards the holdfast.

Texture   Soft but not floppy.

Similars   Many of the frondy red weeds look similar to each other, but *Polysiphonia lanosa* is an epiphyte – that is, it grows on other weeds.

Habitat   Grows on other seaweeds. Common on *Ascophyllum nodosum*, more rare on *Fucus* species and on *Chondrus crispus*.

Distribution   Wherever suitable weeds occur on which to grow, from the Atlantic to the North Sea.

Abundance   Very common.

Picking time   Whenever a food seaweed is picked and *Polysiphonia* is present, eat the *Polysiphonia* as well.

Uses   No specific uses as a separate food species, but can be substituted for other red weeds if wanted.

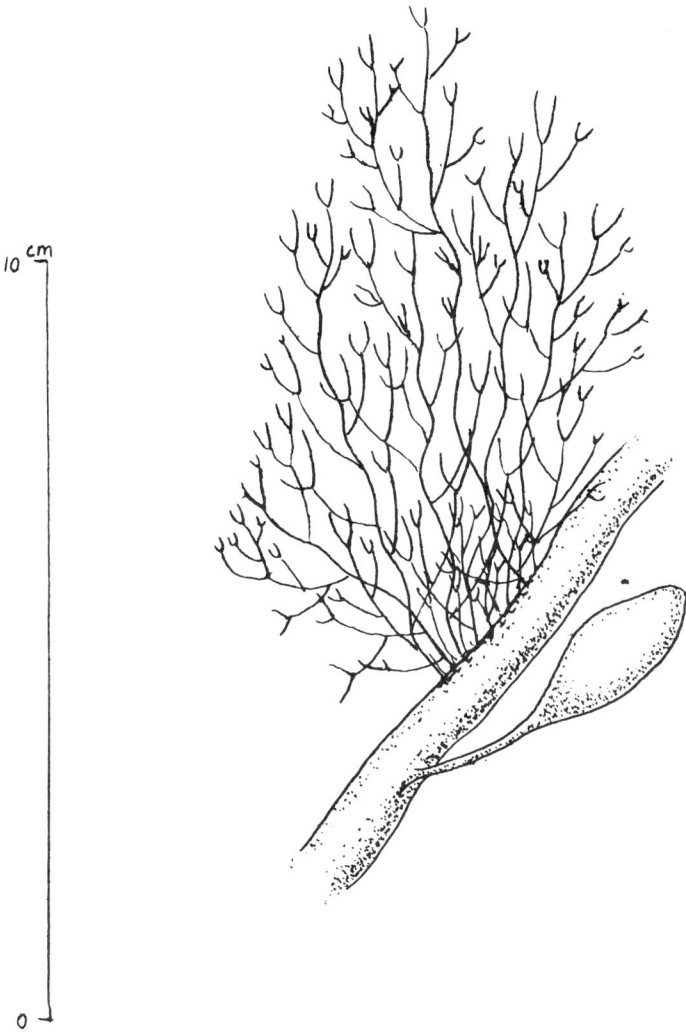

Sargassum muticum
Jap weed.
Not illustrated.

This plant hit the headlines in Britain when it was found in an estuary on the south coast. Now it has spread down to the south west of England, not entirely unaided by humans, and is causing concern among harbour masters who think its great length and ability to colonize muddy areas might mean it would foul the anchorages in the muddy estuaries beloved by small boats. The solution to the problem is to eat it. *Sargassum* could be one of the best slimming aids ever, if it were exploited. Dried weed, eaten as a powder with a drink of water, provides all the nutrients needed by the body, with hardly any calories. Its ability to absorb eight times its own volume of liquid makes the body think it has had a good meal. Because of its very localized distribution at the moment, there is no drawing or description in this book. If you are in the Cornwall or Devon area and find a long strand of light brown weed that looks very similar to the illustrations of other *Sargassum* species in the *Hamlyn Guide to the Seashore*, hold it up in front of you so that the main stem is pulled taut between your hands (it is often more than a metre long). All the side branches should hang down evenly like washing on a line. Check carefully in the identification guide, and if you have any doubt at all take it or send a large dried piece to the Marine Biological Association U.K., The Citadel, Plymouth, Devon, stating exactly where and when you found it, then eat the rest (the dried weed is virtually tasteless).

3

Everyday Uses: Medicinal, Health and Beauty

Seaweed is a valuable source of food, containing protein, fats, carbohydrates, vitamins and minerals. In fact it is almost a complete food, readily assimilated because its K/Na (potassium/sodium) ratio is almost the same as that of the human body. The protein found in seaweed is a 'complete' protein, that is, it contains all the eight essential amino acids as in meat, but has virtually no calories, unlike meat. There are variations in nutrient values between the different species of seaweed, especially fibre content.

	Fibre %	Protein %	Fat %	Carbo-hydrate %
Agar	–		0.3	16.3
Dulse	1.2	20–30	3.2	44.2
Hijiki (*Fucus*)	–	6.0	0.8	30.0
Irish moss	–	–	3.2	–
Kelp (*Laminaria*)	3.0	–	1.0–2.0	50.0
Oarweeds (*Laminaria*)		7.5	1.0	50.0
Green laver	–	34.0	0.6	40.5
Red laver	–	34.0	0.6	40.5
Wakame (*Alaria*)	3.5	12.7	1.5	48.0
Fucus (used in kelp tablets)	6.0	6.0	3.0	55.0

Note: gaps in this table do not indicate a lack of an ingredient, but merely a lack of data.

This table gives just an indication of the value of sea vegetables, but proteins are not everything. It is the other constituents of seaweeds that have justly earned them such a good reputation among nutritionists and gourmets alike, and caught the imagination of the more alert members of our medical profession.

Vitamins and minerals

Everyone has heard of vitamins nowadays, and probably has some idea about what they are and why we need them. Most people then forget about them, and go on through life confident that if they just eat, all the necessary vitamins will be taken in. Unfortunately this is not so. Even the so-called 'balanced diet' is short on vitamins, and minerals as well, for the two are inextricably linked. Modern agricultural methods mean most apples are low in vitamin C, lettuce is hardly worth the effort of eating, and any vegetable grown on a soil deficient in trace elements will itself be devoid of those elements. Nitrate fertilizers and over-liming of heavy land help to hasten the departure of those precious elements needed in minute but vital quantities by our bodies. However, help is close at hand, for seaweed catches and stores those trace elements, and produces vitamins a-plenty so that even the devoutly unadventurous non-seaweed eater can supplement his food intake with a tablet or two, and gain all the benefits of a 'good' diet.

Vitamins do not work independently of each other or of minerals in nature; it seems strange that people expect a synthetically produced 'pure' vitamin to be as good as nature intended, becoming disillusioned and even anti vitamin supplements when the supposed benefit does not occur. For example, vitamin C needs vitamin B and calcium to be present for the body to absorb it properly. There are still so many unknown factors in our food that to depend on artificially produced substances is foolish indeed, and probably even dangerous in the long term.

What do vitamins do?

Vitamin A is soluble in fat and gives a good skin and better vision. Carrots are the usual source, with most vitamin A being lost when food is cooked. Some people cannot tolerate vitamin A supplements, so they should eat foods high in the vitamin instead. Among the seaweeds, the red and green lavers, and various *Fucus* species, have the highest A content.

B complex helps the body balance its cholesterol, repair nerves and improve hair, skin and heart functions.

B6 deficiency causes leg cramps.

B12 is an essential protein food, vital to vegetarians or those on a low protein intake. It also acts as a good pick-you-up, hangover cure or pre-drinking tonic to prevent the ravages of a subsequent hangover. Smoking destroys it, so don't smoke. Vitamin C is the oxygen carrier, fighting infection, and helping to protect cells from outside poisons such as air pollution, smoking, radiation (as does vitamin E).

D is another vitamin soluble in fat, essential for your body to assimilate calcium.

E is the look better, look younger, live longer vitamin. Together with vitamin C it carries oxygen to the muscles. It is generally known as the fertility vitamin, and is most deficient in western diets. Combats fatigue, improves the circulation and is recommended for women going through the menopause.

Vitamin K is involved in the mechanism of blood clotting.

S is an anti-sterility vitamin, one of the more recently named vitamins.

There are probably many more to be discovered and named, which is why a purified supplement of individual vitamins is not recommended. Some vitamins can actually be dangerous if you overdose yourself.

Vitamins in mg per 100g found in some seaweeds (Vitamin A is given in international units)

	B1	B2	B5	B12	C	E	K	
Dulse (*Rhodymenia palmata*)	√	0.63	0.50	1.69	√	24.49	√	t
Hijiki (*Hizikia fusiforme*)	555	0.01	0.02	4.0	√	√	√	t
Kelp	0.33	0.10	0.33	5.7	1.0	13	0.15	t
Kombu (various *Laminaria* plus *Kjellmaniella gyrata*)	450	0.08	0.32	2.0	1.0	11	√	t
Green laver (any large green weed)	950	0.05	0.03	8.0	0.7	10	√	t
Red laver (*Porphyra spp.*)	6000	0.25	1.24	10.0	1.0	22	√	t
Wakame (*Undaria pinnatifida*)	140	0.01	0.02	10.0	0.5	15	0.08	t

t = trace.   √ = present but measurement uncertain.

Vitamin deficiencies are often not serious enough to cause medical concern, and most people tolerate the symptoms as part of their normal way of life. For example, do you have dandruff, dry skin, poor eyesight, bruise easily, thread veins, bleeding gums, anaemia, stunted growth, falling hair or poor fertility? These could all be symptoms of vitamin deficiencies.

Seasonal changes

If you are going to collect your own seaweed, you need to be aware of the changes in concentrations of the various vitamins and minerals in the plants through the growing season. A general rule is that most nutrients are highest in the spring, lowest in autumn, or else the seasonal variation is so small that it really doesn't matter. The graphs on p. 97 of seasonal variations will give you some ideas. Winter pickings can be remarkably good, but be sure to give the plant a chance to grow by leaving the 'stalk and root' intact.

The amount of vitamin B12 present depends not only on season, but on seaweed type. Generally the brown algae are poorest in B12, red algae are rich in it but green are the very best. Folic acid is highest in the green, lowest in the brown algae, whilst niacin and vitamin C are found in about the same quantities whatever the colour of the weed. *Alaria esculenta* seems to have the best counts for all the vitamins.

This seasonal and species variation is one reason why the analyses of various factors in the seaweed differ from the figures of one authority to another. The following table gives a general average of what you should expect to find in half a gramme of kelp (*Laminaria*, and *Macrocystis* species) in a reasonably good season.

The vitamin and mineral content of seaweed

Typical analysis of 500mg of kelp. A conversion table for weights and measures is given in Chapter 8 (µg = microgrammes).

Protein	25–50 mg	Pantothenic	
Fat	10–20 mg	Acid	1.5 µg
Fibre	40 mg	Folic Acid	0.01 µg
Ash	100–150 mg	Folinic Acid	0.01 µg
Moisture	60–75 mg	Biotin	0.05–0.2 µg
Vitamin A	20–33 µg	Vitamin K	trace
Vitamin B1	3–4 µg	Vitamin S	trace
Vitamin B2	3 µg	Aluminium	200 µg
Vitamin B3	trace	Boron	35 µg
Vitamin B5	trace	Bromine	0.5 µg
Vitamin B12		Calcium	6 mg
Vitamin C	250–1000 µg	Chromium	0.5 µg
Vitamin D	2 µg	Chlorine	10–30 mg
Vitamin E	75–150 µg	Cobalt	1.5 µg

Copper	2 µg	Nitrogen	6 mg
Fluoride	trace	Phosphorous	0.5 mg
Germanium	0.25 µg	Potassium	10 mg
Iodine	300–600 µg	Silver	0.25 µg
Iron	5 µg	Sodium	3 mg
Lead	0.5 µg	Strontium	0.5 µg
Magnesium	2–5 mg	Sulphur	10–20 mg
Manganese	20 mg	Selenium	trace
Molybdenum	0.25 µg	Vanadium	0.5 µg
Nickel	1–3 µg	Zinc	25–100 µg

Minerals

Experiments were carried out by G.L. Seifert and H.C. Wood using *Macrocystis pyrifera* (a brown seaweed) as the only source of trace elements in the diets of patients in a maternity unit. Patients who were anaemic, subject to colds and the euphemistically termed 4-o'clock fatigue all showed a change for the better. Even those that actually caught colds did not suffer unduly from the virus. The miscarriage rate was well below that of the general population. There was a general improvement in body metabolism, no constipation problems, good digestion, etc.

What do minerals do?

There is still a lot to learn about the body's need for and use of trace elements. How much each person needs depends entirely upon that individual, so maximum or minimum daily limits are almost meaningless. That we all need 'some' is certain, but absence of one or two minerals may not reveal itself in readily identifiable medical symptoms. As with a shortage of vitamins, the results may just be a lowering of vitality, less resistance to common infections and a reduction in the quality of life that can lead on to more specific and serious problems later in life. Most of the data we have on minerals comes from American work, the British and Europeans lagging far behind in this type of research. The following list is a very brief resumé of the commonest and most plentiful minerals present in seaweeds and what they are used for in the human body. However you are recommended to read one or more of the excellent books on minerals now in the bookshops, and likewise for vitamins.

Calcium is well known and well researched, and serves as a good example of the interdependence of vitamins and minerals in your body. In 500 mg of kelp there is approximately 6 mg of calcium, or Ca, to give it its scientific initials. Bones and teeth use 99% of the calcium we take in. The human skeleton is constantly renewed, in the adult one complete change taking about ten years. This could be one of the reasons why an intake of noxious substances does not cause cancer until many years later. Although calcium is the main raw material, it needs other elements present and good nutrition generally in order to work properly. The odd 1% is used in the blood, muscles, nerves and various enzymes' activities. It is essential for muscle contraction, and in the reaction of nerves to stimulation. Calcium is needed in blood clotting, and to trigger off the enzymes required by other body processes, particularly to help us use our intake of iron and vitamin D, and also to regulate the functioning of hormones.

The human body finds it difficult to absorb calcium, so it needs a 'carrier', which just happens to be vitamin D, and a little but not too much protein. Kelp has vitamin D, protein and calcium all together. Foods containing oxalic acid (tea, cocoa, chocolate) and saturated fats (eggs, cheese) will impair calcium absorption. So remember, milky tea is not much use except as a thirst quencher. Finally, it is interesting to note that the body loses calcium daily, more if you are confined to bed or just not exercising. A brisk walk to the beach to gather your own kelp would seem to be the ideal way of getting and keeping your calcium balance right.

Chromium helps your body cope with the sugar intake in your food, though how this comes about is not fully understood. Experiments on the poor old rat showed that if it were deprived of chromium it developed symptoms similar to diabetes, had arteries clogged with fat, and a high level of cholesterol in its blood. This is one of those trace elements that is lost from the land under modern intensive farming methods, and therefore is very unlikely to be present in your nice fresh vegetables. But in 500 mg of kelp there is 0.5 µg of chromium.

Cobalt is another 'unknown'. Too much can result in goitre, or 'Derbyshire neck', because it interferes with the thyroid gland. But a little is absolutely essential. Vitamin B12 is vital to human survival. Our bodies cannot manufacture this vitamin, so we usually obtain it from meat. For vegans (who do not eat dairy produce), shortage of B12 can be a serious problem. Cobalt is part of the B12 molecule, and is also linked to the manufacture of the thyroid hormone, which requires about 0.4 microgrammes (µg) of cobalt per day. Amounts required by the other body functions are as yet unknown. There are 1.5 µg of cobalt in 500 mg of kelp.

Copper. Small amounts are needed in the formation of melanin pigment (brown-ness of skin), red blood cells, and the RNA (ribose nucleic acid) which carries the genetic coding of each individual. The mineral is also used in bone growth, and helping the body cope with its intake of iron. Lack of copper occurs where there is too much molybdenum in the soil and/or too much fluorine in the water. If babies are weaned too early onto cow's milk and cereal, they may suffer a copper shortage. People on penicillamines, who take oral contraceptives, or who live on too much processed food also run the risk of deficiency. In animals, the shortage results in diarrhoea, loss of hair colour and even loss of hair. The copper bracelet worn by many sufferers from rheumatoid arthritis, and which often gives relief, may be an indication of their need for copper intake. In Hastings, Sussex, I can remember seaweed baths being very popular among 'rheumatic' people, and an elderly aunt would spend her entire holiday with us steeped up to her neck in warm seaweed. The treatment worked like a dream and she would go away revitalized for the year ahead.

Fluoride has become a controversial mineral in the past few years. This waste product is dumped in our drinking water on the ill-founded assumption that it stops tooth decay in sugar-addicted children. Most of the fluoride in our diets comes from tea drinking and seafood, although traces can be found in every food. If your drinking water contains 1 part per million of fluoride (check with your water board), then drinking 1¾ pints (0.84 litres) per day will give you an intake of 1 mg. Add to that 0.33 mg from each cup of medium strength tea, that is, from the tea alone, and your intake soon mounts up. Here is the sting in the tail: too much fluoride causes mottling of the teeth and *decay*; too much is only 5 mg a day or over. Overdoses of fluoride can cause stunting of growth in plants. No one has studied its effect on people, but to date no researcher has found any therapeutic use for fluoride. It is one of those mysterious minerals we can almost do without, and in kelp there is just a trace.

Iodine hits the news whenever there is a radiation scare.

Although its importance for the body is known, how much we need is still a mystery. Some authorities give ranges of 50 to 300 µg daily, and kelp is the best source of this, having 300 to 600 µg per 500 ml (dried). The thyroid gland produces hormones which control the whole bodily rate of activity, either directly or indirectly, and it needs iodine. Most of us have come across the two disturbances of thyroid which are so obvious; the very fat person whose whole metabolic system has gone wrong due to a 'slow' thyroid activity; or the hyperactive person, thin as a rail and with staring eyes, who has an overactive gland. More rare nowdays is goitre, once common in hard water areas where the iodine count is low. Too much eating of cooked cabbage can interfere with the absorption of iodine, whereas raw cabbage, being difficult to digest, does not interfere so much with iodine absorption. In Japan, where cabbage is eaten raw, if at all, iodine deficiency is very rare. The Japanese eat lots of seaweed, which rectifies any iodine deficiency.

Iron is the best known of minerals. Everyone knows that insufficient iron causes anaemia. But there are some strange fallacies promulgated: adding iron to commercial foods has become an important advertising point, but, in truth, the iron is in a form which the body cannot readily assimilate. Also, iron being an 'unwilling' subject for digestion, you must have vitamin C and copper to help in its absorption. Kelp and other seaweeds provide all the necessary vitamins and minerals to help absorption, as well as the iron itself which is in a more absorbable 'natural' form as well. In 500 mg of kelp there is about 5 µg of iron, but it is all useable. Recommended daily intakes range from 6 mg for babies, 7 to 15 mg for ages 1 to 17 years proportionately, 10 mg for men, 12 mg for women, 15 mg (U.K.) or 18 mg (U.S.A.) for pregnant women. Before swallowing dozens of kelp pills check up in various books on the iron content of other foods you eat, as an excess of iron can cause digestive disturbance.

Potassium is vital to the proper functioning of nerves, muscles and cells. It is needed by cells to help maintain their osmotic balance, that is, keep them in balance with the blood and body fluids around them, not too plump and watery, not too concentrated or shrunken. There is a sodium/potassium 'pump' which operates within the body's cells, across the cell membranes. Quite an ingenious device. Potassium is also used in the conversion of glucose to muscle energy. An average balanced diet should produce 3 to 5 g per day, but, as we all know, the ideal balanced diet just does not exist. Potassium is lost if you use too much common or table salt, sodium chloride. Your body has to work hard to get rid of the surplus sodium and so loses potassium at the same time. Any drug that increases elimination from the body such as enemas or diuretic tablets will have the same effect, causing loss of a valuable element. Too great a loss results in weak muscles, mental confusion (very common in the elderly) and heart failure. Processed food with all its added salt is generally low in potassium. Kelp contains 10 mg in 500 mg of the weed, and is a good salt substitute as well.

Selenium is a trace element that has lately become fashionable. People in areas where selenium levels in the soil and water are high have shown lower heart disease and cancer rates. Too much of a good thing can cause the 'staggers' in animals, but a little selenium prevents some vitamin E deficiency. Together vitamin E and selenium appear to protect cells from breakdown, and stop destruction by heavy metals. In particular, smokers need a high intake to counteract the cadmium in tobacco smoke. (See p. 86, radioactivity and metals.) Deficiency can cause failure of the heart, liver and kidneys, plus sterility, capillary breakdown and poor blood circulation. A definite sign of low selenium levels is dandruff. Long cooking or processing of food results in loss of the element. In 500 mg of kelp there is a trace of selenium.

Sulphur, selenium and zinc are linked together in the maintenance of healthy skin, hair and nails. 500 mg of kelp contains 10 to 20 mg of sulphur.

Sodium is the salty part of salt, sodium chloride. Our bodies need a little, but the daily intake in our (unprocessed) food is quite enough. Salt added to food is overdosing the system, and thus causes the loss of other valuable minerals such as potassium. It can also raise blood pressure too high, and cause heart trouble. Modern low sodium salts are now popular but better to use kelp powder if you want a salt substitute. Its 3 mg of sodium is in balance with all the other minerals and vitamins present.

Strontium is newsworthy when radioactive, and in milk, but it is part of our bodies all the time. It is always found with calcium, and behaves similarly. However, little is known about its effects or its deficiencies. 500 mg of kelp contains about 0.5 µg of strontium.

Vanadium is an enigma. Most research so far has concentrated on the effects of deficiencies in animals and the results have been unpleasant, with shortened lifespans, reproduction rate down, retarded growth and raised cholesterol levels. Like most other trace elements, vanadium is lost in food processing. In 500 mg of kelp there is 0.5 µg of vanadium.

Zinc has now been recognized as a vitally important element in man's diet. It is necessary for normal growth, both in man and in the plants and animals he relies on for food. It is the essential ingredient for many enzymes, especially insulin. Enzymes are the chemicals that break down our food to a point where the body cells can utilize it. Zinc is particularly important for pregnant women. A lack of this element is indicated by slow healing of wounds, white spots on fingernails, lack of sense of taste, rough skin, poor growth of hair, prostate trouble and ageing. In the more extreme form there can be lack of sexual development and dwarfism, arteriosclerosis, cancer and diabetes. Processed food, again, lacks zinc, as does water passed through copper piping. Heavy drinkers lose the element, and women on oral contraceptives should take a supplement. The estimated daily need of zinc is 15 to 20 mg, but kelp provides 25 to 100 µg per 500 mg.

Although kelp has been used as the standard reference so far, because it is the seaweed on which most research has been done, some information is available for the principal mineral concentrations in other seaweeds.

The whole complex of minerals is interdependent; therefore it is reasonable to suppose that a purified extract or synthetic substitute of one mineral alone is of no use at all, and may even result in an adverse or overdosed reaction. Unfortunately it is a modern demand that all results must be instant. Better health from eating seaweed, or taking seaweed tablets as a supplement to a good diet, will only come relatively slowly. But the results will be marvellous and long lasting, and your patience will be well rewarded.

Radioactivity and metals
Although we need traces of metals to keep our bodies functioning, in modern everyday life we are bombarded with overdoses of

Principal minerals (mg per 500 mg of weed unless otherwise stated)

	Calcium	Iron	Iodine	Magnesium	Phosphate	Potassium	Sodium
Agar	2.8	0.03	0.1µg	–	0.11	–	–
Dulse	1.5	0.75	0.04	1.1	1.34	40.3	10.5
Hijiki	7.0	0.15	–	–	0.28	–	–
Irish moss	4.4	0.04	–	–	0.79	14.2	14.5
Kelp	5.5	0.5	0.75	3.8	1.30	26.4	15.0
Kombu	4.0	–	0.38	–	0.75	–	–
Nori green and red	2.4	1.15	–	–	2.9	–	–
Wakame	6.5	–	0.04	–	1.3	–	5.5

Note: the gaps mean no analysis done or insufficient data available, not that these minerals are absent from the seaweed.

nasties such as lead, and increased radiation levels. Seaweed is useful as a cleansing agent in the body. In December 1965 the scientific magazine *Nature* published a report showing that kelp inhibited the body's absorption of radioactive strontium and cadmium, up to seven-eighths the radioactive dose received. It also removed Strontium 90 which had already been absorbed into the body tissues.

Radiostrontium goes through the gut wall and is deposited in the bone. This can cause bone tumours and other changes. If Strontium 90 is in the bones already, a diet rich in alginates will remove it. The best way to take alginate is as calcium alginate if you anticipate a long period of treatment, so the calcium balance in the body will not be disturbed. There is no chemically pure form of alginate yet, although the main active ingredients are known, being guluronic and mannuronic acids. But there are other, unknown, factors which control the effectiveness.

Toxic metal waste is removed from the body by agar. It acts like pectin (found in apples) to bond with the metals such as lead. Alginates, from brown seaweeds, bind to barium, cadmium and zinc. An added advantage is that alginates are non toxic and cheap to produce.

The *Fucaceae* could be of use in illnesses such as siderosis, where the body is absorbing too much iron, and in alcoholism, because of their affinity for ferrous iron, binding it and removing it from the body.

It may seem strange to be singing the praises of seaweed as a valuable source of minerals, then saying how it can be used to remove metals from the body. This is the paradox of a natural system, working both ways at the same time, removing imbalances, restoring things to the way they should be. Something that no modern wonder drug has ever managed to achieve.

The following table shows what percentage inhibition of a radiation dose various seaweeds can accomplish. Of course, you do not pick your seaweed near the outfall from a nuclear power station.

Most effective percentage inhibition of SR89 (Radiostrontium)

Macrocystis pyrifera	80
Egregia menjiensi	74
Nereocystis lusetkeana	69
Alaria marginata (dabberlocks)	60
Laminaria digitata (oarweed – brown)	60

Generally seaweed is a tonic, building up the body's defences, improving health and well-being. Through the ages certain weeds have been sought out for their action on specific complaints, and the results of eating sea vegetables over a long period of time have been noted. The following list is just the tip of a great iceberg of knowledge from all over the world. It shows the medical uses of various seaweeds.

Aonori
Chlorophyceae, or green seaweed. Species of *Ulva*, *Monostroma* and *Enteromorpha*. A source of salt. May be eaten raw or cooked. The Chinese use it as a poultice to cure warts and haemorrhoids, and internally for stomach ailments.

Arame (*Eisenia bicyclis*)
Phaeophyceae, or brown seaweed. In Japan it is a species called *Eisenia bicyclis*, but the nearest in our cool-temperate waters are *Laminaria hyperborea*, *L. digitata* and *Saccorhiza polyschides*. It is useful as an emergency food, keeping its taste even after being dried for two or three years. Medicinal uses: feminine disorders,

problems in the mouth such as sores and ulcers, high blood pressure. This could be because of its high concentrations of calcium and potassium.

Bladderwrack (*Fucus vesiculosus*)
Another brown seaweed, known as sea wrack, pop weed, cut weed or rockweed. It has a strong smell and taste of seaweed, and was formerly used for iodine production. The iodine in the weed is both a hormone and an antibiotic. A simple use is as a suntanning preparation; just rub the weed on the skin. Roast bladderwrack can be mixed with a plain oil such as olive oil to give a good alternative to cod-liver oil. For those suffering from obesity, 1 oz of dried weed to 1 pint of boiled water, taken in a wineglass or better still in pill form as kelp tablets, will have a good effect on the kidneys, and stimulate digestion and clearance of the system. It also hastens the renewal of tissues.

Dulse
Rhodophyceae, or red seaweed. A most popular food, with the highest concentration of iron of any food source, plus potassium and magnesium. It makes an inexpensive and pleasant flavoured substitute for chewing tobacco and the modern day potato crisp. But its main use is as an antiscorbutic (source of vitamin C – used to cure scurvy).

Hijiki (*Hizikia fusiforme*)
Another brown weed, of the kelp family and related more to the bladderwrack than to the large blade kelps. Its greatest medicinal use is in cases of overweight, where it is useful as a food: the dried seaweed can absorb water and expand to four or five times its volume. Add to this its content of vitamins and minerals, and virtually zero calories, and you have an ideal slimming food.

Irish moss (*Chondrus crispus*)
Red weed, but tougher than other weeds, with a higher sulphur content. Whole plants used to be tossed into brewery vats where they would bond with impurities and carry them to the bottom. A modern medicinal use is as an anticoagulant with its own built-in time-release factor. Age-old remedies were for chronic coughs and bronchitis, bladder infection and kidney irritation, and as a medicine to soothe and protect the intestinal walls. A decoction

was made by soaking ½ oz in cold water for 10 minutes, then boiling in 3 pints of water for ¼ hour, straining and flavouring.

Kelp

A broad description of most of the largest brown seaweeds round our coasts. They are harvested on a commercial scale in the USA and go to produce algin (the gluten or sticky substance in seaweed) and kelp products such as tablets and powder, a valuable source of minerals and trace elements. The kelps are used for arthritis treatment, to help clear sterility, for prostate trouble, and ovaries that won't work properly. The natural sugars, fucose and mannitol, make it ideal for diabetics. Kelp powder is also a good salt substitute. Peruvians and Sherpas on opposite sides of the world carry little bags of kelp which they eat when at high altitudes, to aid breathing and restore tired leg muscles.

Kombu

Another kelp or *Laminaria*. It is said to be the great balancer and almost the complete food. It aids digestion, cleans out the alimentary system of debris and unwanted food residues, stimulates the tissues to greater activity, relieves anaemia, normalizes the reproductive organs, stimulates sluggish kidneys, in fact, generally makes the whole body sit up and take notice.

Laver or nori

Whether in English (laver) or Japanese (nori), the red *Porphyra* is a favourite food. It aids digestion and decreases cholesterol, relieves acute bouts of gallstones, stimulates the liver, and has been variously recommended for the cure of warts and rickets. The green weeds have been shown to have the best effect on cholesterol levels (*see* Aonori).

MEDICINAL USES

Seaweed contains so many good things that it is not surprising there should be many medical uses for it. Modern pharmaceutical industries use agars and alginates in pills, lotions, potions and pastilles. Alginates are especially useful where a medicament has to get through the stomach untouched and do its work in the intestines. Stomach acid is strong, but it makes little impression on that particular jelly coat. However, the old remedies have had a few thousand years of testing, and in many parts of the world are still the favourite cures. It is even better when a favourite food is also a favourite medicine.

As an antibiotic, seaweed is best in the spring. Even if the material is oven dried or freeze dried, the activity is the same. Different weeds work best on different bacteria, but *Polysiphonia fastigiata* shows the widest bacterial spectrum, that is, it kills off more types than do other seaweeds.

Ageing

The Japanese consider that seaweed as a food counteracts the effects of ageing. Cerebral haemorrhage and high blood pressure are rare in those people who have a regular intake of sea vegetables. This could be in part due to the vitamin K content, which is an anti-haemorrhagic. Zinc deficiency stops the body's cells from 'breathing' properly or efficiently, so the high quantities in kelp will aid the system, and improve cell activity. This in itself should act like rejuvenation.

Anaemia

Vitamin B12, found in all seaweeds and especially in kelp, should help this condition if it has been brought about by insufficient intake of iron (iron is not always easily absorbable). There is also iron in seaweeds, for your body to take what it needs.

Arthritis

Arthritis is a mineral-related affliction, where disturbances in the acid balance of the blood and lymph are the strongest contributors. Kelp has a potassium/nitrogen ratio of 3:1, which is close to the human body's ratio of 5:1. The plant is very useful in restoring the body's balance, and has been used to treat arthritis, rheumatic fever damage and heart pains.

Bladder

Irish moss has been used to soothe irritating bladder disease, taken as a decoction when needed. Bladderwrack, despite its name, has only an indirect effect, in that it influences the kidneys, and again restores balance in the body.

Bronchitis and chronic cough

A good remedy is Irish moss, ½ oz soaked in cold water for 10 minutes, boiled in 3 pints of water for ¼ hour, strained and flavoured. Kelp, the universal remedy, will aid breathing.

Cholesterol
Reduce it by eating the red *Porphyra*, as laver bread or nori, or better still the green aonori.

Depression
A strict seaweed-only diet, based on kombu, to rejuvenate the whole body.

Constipation
Agar agar used as a jelly, flavoured as you wish. The pharmaceutical industry favours orange in its commercial preparations.

Female disorders
Phaeophyceae (brown seaweed), such as *Laminaria* or Japanese *Eisenia* are rich in potassium and calcium. Taken as a food, or as a tea, to help overcome nerve and muscle problems.

Haemorrhoids
Internally, take *Chlorophyceae* as a food. Externally, use as a poultice.

High blood pressure
Brown algae, with its calcium and potassium, encourages you to eat or drink the weed and be merry.

Mouth ulcers
A tea of brown weed such as *Laminaria saccharina* washed around the mouth several times a day will heal and soothe.

Obesity
A remedy for obesity is the following: bladderwrack, 1 oz of dried weed to 1 pint boiling water; a wineglassful three times a day stimulates the kidneys. Hijiki as a food absorbs five times its volume of water, with almost zero calories, so this is a good aid also.

Radiation exposure
See the beginning of this section (p. 87).

Sterility
In men, kelp helps to clear up prostate troubles. In women, kelp regularizes the functioning of the ovaries.

Stomach ailments

Green seaweeds as food are easily digested, whether raw or cooked. A way of taking in valuable vitamins and minerals with the least irritation.

Warts

Put on a poultice of green seaweed. An old Chinese speciality.

BEAUTY: FOR WOMEN AND MEN

Thick glossy hair, a clear complexion, pliant skin that is free of wrinkles – a dream? Not so if you eat seaweed every day and use it in your beauty preparations. The Japanese women were famous for their beautiful hair and skin before the western diet corrupted them. Some Japanese firmly believe that the Americans invented acne, muddy skins and yellow eyes. A little hijiki in your meal will normalize blood sugar levels, and give a radiant complexion. Vitamins and minerals will feed the cells, encouraging renewal, getting rid of accumulated waste. The natural antibiotics found in seaweeds will clear up minor infections and promote a feeling of well-being. All seaweeds act as tonics, and once you eat or take them regularly you should not need any artificial boosts. For example, Irish moss helps maintain the glandular balance and its calcium chloride content is a gentle heart tonic.

Most of us need to lose weight, so here are three diets based on seaweeds. These diets are not just seven-day wonders, but the way into a new habit, a new way of feeding. Don't just diet to lose weight, instead tell yourself your eating habits are now changing for the better, and for all time, so you can feel fitter, live longer, and so on. Complete recipes will be found in the appropriate chapter, and I have tried to calorie count each one for those who like to keep to a daily limit.

First, the easy one

This is the slowest, easiest, laziest way to reduce. You will need lecithin capsules, kelp tablets with some cider vinegar, all available from your local health store and, nowadays, most chemists as well. You may come across 'omnibus' pills containing all these ingredients, sold as 'fat burners', but the basic idea of the cider vinegar is to re-educate your taste buds as well as burn up the calories. So I would recommend you put up with the bit more fuss involved, and go for the separate items.

First thing in the morning, take one tablespoon of cider vinegar in a glass of warm water, together with the amount of lecithin and kelp recommended by whichever brand you have bought. Then you can eat, but try to be sensible. Remember, when calorie counting, fat is worse than sugar, but sugar is bad enough. From now on, half an hour before every meal take the cider vinegar, lecithin capsules and kelp tablets. After the first week try increasing your vinegar dose to two tablespoons. Over a period of months there should be a decrease in weight and an increase in vitality, but your eating habits and conscience will decide the issue.

Second, the sensible one

Keep to a calorie-controlled diet if you like doing sums; 1,000 a day is the recommended level. To fill an empty tum, eat fruit-flavoured agar jelly (see recipe section, p. 143). Use agar for thickening soups, sauces and to replace egg in flans. Take kelp tablets to ensure you don't run short of those vital elements. Have seaweed in at least one meal each day. Cut out most fats and oils, most sugar, some carbohydrate, some protein. Watch out for eggs, they are good in minute amounts, but best left alone. Eat when you feel like it, not according to someone else's clock, but try to get the last meal in before 7.30 p.m. After that your body would rather rest than digest and so tends to store imbibed goodies against the famine of the next day.

There are various helpers in the form of 'fill-up pills' in the shops, but read the labels carefully. Meadow Croft produce a fibre formula which contains among other things lecithin, kelp, cider vinegar and *spirulina*.

Third, the kill or cure

The easiest part is the first three days. No food, just water with a dash of cider vinegar if you like it, but not too much. Then use agar, *spirulina* (at end of recipe section), kelp and other seaweeds to provide your main food intake. They should be used as a complete food, filling, yet very low in calories. If you feel your will weakening, then have one meal of no more than 500 calories. Weight loss will be rapid, but be sure to keep up your liquid intake. Don't over-exert yourself, but do take some form of constant gentle exercise. Swimming is ideal.

For a one-week detoxification, take *spirulina* and fruit juice, nothing more, nothing less.

Having now dealt with the inner person, let us have a look on the outside. A clear and radiant face should have no need of make-up, but if you desire a little rouge to brighten up a winter aspect, or for a party, then follow the example of the Roman women. They used to beautify their faces with a cosmetic made from *Fucus linnaeus*. The women of KarnChatka mix their *Fucus* rouge with a little fish oil, but I think for western taste a touch of lanolin or some light cosmetic oil would be more acceptable. For real luxury, use almond oil, or just dust lightly over a foundation cream. Half the fun comes in experimenting.

Face pack
Spread *Ulva lactuca* (sea lettuce) on a cleansed and warm skin, wrap face in warm towels and relax till cool. Rinse off with warm water, then splash face alternately with cold and warm water, ending with cold. Pat dry.

Or, take a tablespoonful of each of the following: Fuller's earth, honey, milk, mashed or liquidized red and green seaweeds. Add more Fuller's earth until the pack is thick enough to spread on the skin. It looks horrible, but skin loves it.

Your face and body have one thing in common: skin. Seaweed extract can be stimulating and refreshing. There is no need to submerge yourself in a tub of warm and slippery weeds any more, Lincoln Fraser products now produce a seaweed mineral bath (address at the end of the book). Dickens and Jones sell a seaweed body treatment, massage cream, soap and cream bath. Some beauticians will give you the seaweed treatment if you ask loudly; a little publicity could make the products available more widely. But there is nothing to stop you making your own.

Suntan lotion
Glycerine 1 part, alcohol (use vodka) 5 parts, seaweed water 40 parts. Seaweed water is made by steeping ½ lb dried bladder-wrack in 3 pints water, then boiling for 5 minutes.

Suntan oil
Sesame or olive oil 1 cup, cider vinegar ½ cup, pulverized bladderwrack 1 tablespoon, few drops of lavender oil to keep the insects away.

Mineral or seaweed bath
Magnesium sulphate 100 parts, powdered or fresh, chopped or

liquidized mixed seaweeds (get a good variety) about 20 parts, a few drops of perfume oil to mask the seaweedy smell. Pine is good, if you like it.

Shampoo
Seaweed shampoo for the hair is hard to find. Creightons of Sussex produce a nice one, not at all weedy smelling. Usually a *Chlorophyceae* or green weed is used. The chlorophyll is a natural deodorant, iodine acts as a germicide, and the colour is natural. Ideally, seaweed shampoo should be used on normal to oily hair. It has an emulsifying action which is a bit too drastic for fine dry hair. The Body Shop does a lovely seaweed and birch shampoo. If your hair is thinning, try eating more hijiki and wakame.

Oil-based shampoo, to get rid of dandruff
Castor oil 5 parts, olive oil 5 parts, kelp powder 1 part, water 30 parts, a few drops of perfume. Rub into hair prior to shampooing out with a baby shampoo. The hot towel treatment helps the effect. If you add sage leaves crushed, it helps to darken grey hair.

Tooth powder
For your teeth, make a tooth powder of powdered carragheenan (the jelly produced from carragheen) 5 parts, calcium carbonate 10 parts, a good pinch of sea salt, a good pinch of powdered green seaweed, a good pinch of rubbed sage leaves, a few drops of peppermint if you like. Keep in an airtight tin and apply with a damp toothbrush.

Toothpaste of a sort can be made with glycerine 2 parts, chalk 4 parts, Irish moss 1 part and a pinch of sage.

VITAMIN PEAKS IN *Alaria esculenta*

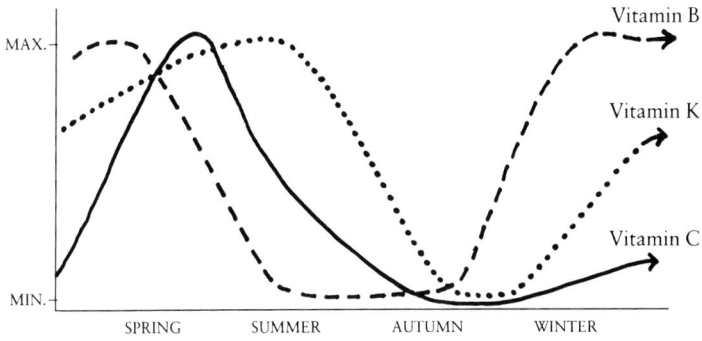

SEASONAL VARIATIONS IN THE CONCENTRATIONS OF
VARIOUS NUTRIENTS IN SEAWEEDS IN GENERAL

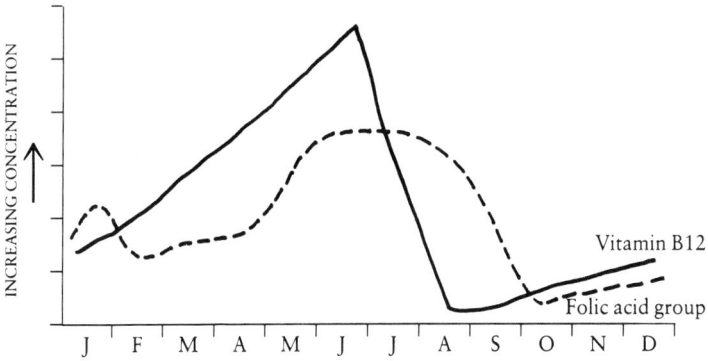

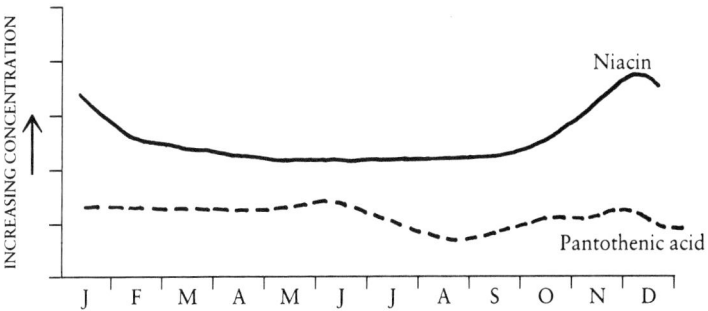

4

Everyday Uses: for Animals

Although seaweed has been used in western Europe as an animal food for thousands of years, in the eastern countries it was rarely used for this purpose, being thought of as a food source for humans only. Early records show the Romans using seaweeds as part of the rations for horses. No one notes whether the horsemen themselves ever had a crafty nibble of sea lettuce, or a dulse dinner. Modern times have changed many of the old practices, though in the Scottish Islands the sheep still graze the sea wrack, and sometimes in winter seaweeds are boiled up, mixed with oats, husks and hay for animal feed. But most of the sea product nowadays is processed to powder form, so that factories can mix it into commercial feedstuffs.

Of all the seaweeds available, the commonest, *Ascophyllum nodosum*, is the favourite of the commercial producer of animal feeding stuffs. The chapter on human medical and health uses gives data on the vitamin, mineral and other constituents of kelp, so in this section, only data for *A. nodosum* will be used. Because of its importance as a feeding stuff, the chemistry of *Ascophyllum* has been thoroughly investigated. It does not change much in composition throughout the seasons, and when dried and processed the amount of nutrients lost is insignificant, indeed vitamin A stores better in it than in dried grasses. *Ascophyllum* plants are dried to no more than 15% moisture content, the calorie value is low, and besides all the usual vitamins, minerals and trace elements, carbohydrate content is from 40% to 60% and protein is between 5% and 10%.

If seaweed meal is combined with other fodder sources such as roots and hay, it makes a good winter feed for animals. The commercial preparations available are carefully balanced and expensive. They take no account of the mutual supportive action that vitamins and minerals give, because such action can rarely be measured in a test tube. So amounts are calculated on a straight content of each item. Herbivorous animals prefer to graze on seaweed if given the chance, and take many varieties. Common names of some seaweeds indicate their popular uses. *Ascophyllum nodosum* is known as pig weed, *Alaria esculenta* is cow weed, *Rhodymenia palmata* is cow or horse weed, depending on whether you are in Brittany or Norway. Commercial seaweed meals based on just one species are only second best; better to gather a great variety as the season offers opportunity and store for future use. Drying the weed is a problem, especially on a large scale: perhaps try a windmill-driven drum dryer, or hot air fed in from a small furnace that is used for other things as well. Heat is always going to waste in the average house, much better to use it drying seaweed.

Many experiments have been carried out to try and show how animals utilize the nutrients in seaweed meal. Farmers, using rule of thumb and knowledge of their animals' previous conditions, say that a seaweed diet for all or part of the year gives sleek coats, good health and wards off disease.

Results of trace element deficiencies are well documented. A shortage of copper results in anaemia; too little cobalt gives 'pining', coast disease and marasmus; zinc deficiency makes for soft egg shells and hair loss, and so on; any farm vet book will give you the facts. French farmers believe that calcareous red seaweed (maerl) fed to livestock prevents tuberculosis and foot-and-mouth disease. A healthy animal will resist illness better than one that is a little 'under the weather'. Vitamins and minerals present in seaweed, although in theory less than the required daily dose, have a synergistic effect upon each other. This mutual self-support boosts the value to the animal, and probably increases absorption and utilization of food content. The problem is how to measure such an action in order to prove the theory, at least that is the problem for the scientist. Meanwhile the dedicated seaweed-fed chicken lays strong-shelled eggs with good yellow yolks; milk yields from cows, sheep and goats increase; pigs put on less fat and more meat, and, as an extra bonus, don't feel the cold so much. Pigs also resist parasites better, and the seaweed keeps them 'regular', resulting in glossy coats and a healthy skin. Sheep grow thicker winter wool with a longer staple. Where scientific experiments wered conducted on 3,500 sheep in Norway, there was an increase of 20% in winter wool production, due mainly to the seaweed food supplement preventing winter moulting. The loss of lambs from white muscle disease was also reduced significantly, and generally lambs grew well.

Cows benefit from seaweed added to their usual rations. The vitamin E content ensures that reproductive performance is improved. Again in Norway, a long-term experiment showed that in seaweed-supplemented dairy cows, the number of services per conception was down, in many cases one service was enough to produce pregnancy, and the frequency of mastitis dropped dramatically. Improved milk yields could be due to the weed feeding bacteria in the cow's stomach, so improving food break-down and utilization. Iodine deficiency in cattle causes still-births, and some feedstuffs such as kale and swedes may make it more difficult for animals to absorb iodine. Seaweed balances the diet, putting in iodine with all its complementary vitamins and minerals, and increasing absorption by the body.

Chickens breed better when seaweed meal forms part of their diet, and chicks on 5% meal in their chick crumbs grow fat, fast and fine feathered, especially if yeast is added as well. In horses, addition of seaweed to the feed reduces the incidence of in-flammation in tendons and nerve sheaths, prevents cracked hooves, and increases the fertility of brood mares.

Wherever seaweed is used, iodine will be found in the milk or egg yolks. Therefore using dried *Laminaria* as part of the feed is not a good idea because of the very high iodine content. Howev-er, if your animals are free grazing on fresh wet weed then do not worry. Besides having an innate sense of what is best to eat, animals will take in so much water contained in the plants that the more 'powerful' constituents will be diluted.

Much has been written over the years about the problems and dangers of adding pure chemical extracts to food. Just because a manufacturer states that animals need certain minerals and his pure feed provides them all, does not mean he is correct, or that the animals can actually utilize the additives. No single vitamin or mineral works entirely alone, or in the pure form. Indeed the chemically derived minerals may prove to be poisonous in quite small quantities. There are many trace elements in our foods which research workers are only just beginning to recognize as having possible importance, and even more than that are prob-ably still unknown.

Mineral and iodine deficiencies in animals can be remedied by seaweed. Feeding weed to animals and humans returns the valuable nutrients to the land, ending the cycle of mineral loss due to natural erosion and poor farming practice to the benefit of all living things.

5

Everyday Uses: on the Land

In coastal agricultural regions of the world, seaweed was used for fertilizer in historic times. Unfortunately the everyday practices of the small farmer and peasant are rarely recorded, and it is only from the reports of travellers and writers that we can discover traditional country practices; the sandy soils of Scotland have benefited from additions of seaweed since before the 1500s, and until the 1960s farms of the Channel Islands, Scillies and Cornwall spread the sea harvest on their fields, then ploughed it in. Production was high, disease minimal. The coastal areas of Brittany and Normandy were known as the 'Golden Belt' of France because of rich harvests from seaweed-fed land. Cheap chemically pure fertilizers have replaced the natural weed; production is high, but now the cost per acre is rising fast.

Seaweed presents a problem to the farmer because it has to be cut by hand, then brought to the farm. Modern agricultural holdings exist with as few workers as possible, but much machinery. Time is precious, and cannot be spared to cull a free harvest. Donkeys no longer toil up steep lanes loaded with wet weed and sand. Farm workers cannot do anything at a slow pace any more, so the old way has had to die out. With interest rising in organically grown vegetables, modern commercially produced seaweed meals and liquids look set to come into their own.

But what effect does seaweed have on the soil and on the plants grown? Carbohydrates in the plant rot down and encourage soil bacteria to multiply. Their soil conditioning action improves the structure, making a good 'crumb' aggregate which remains stable. This improves the soil's ability to hold water. At the same time nutrients are released into the soil in forms which plants can easily use. Trace elements held in the soil in an inorganic form, unusable by the plants, may be released and made available to those plants when seaweed is added to the soil. Although any rotting plant material reduces available nitrogen, seaweed breakdown has the least effect because it contains so little cellulose.

Even so, using whole weed on farmland should be looked on as a long term fertilizer, the gradual breakdown and release of nutrients into the soil ensuring that good plant growth is maintained. It is rare for the small amount of common salt contained in seaweed to have any adverse effect on plants, in fact sugar beet and asparagus love a little added salt. But if you wish to use whole wet weed as a fertilizer and are doubtful about the sodium chloride content, then spread the plants on the soil surface in the winter, dig in in spring. Where a light sandy soil is deficient in potash, seaweed is particularly good. It is a perfect manure for barley growing land, especially if a little seaweed is put into the bottom of seed drills to conserve moisture.

To sum up, the advantages of fresh seaweed used as manure are threefold: it contains a wider range of trace elements than farmyard manure; it improves the condition and structure of the soil; it is free of weed seeds and fungal spores which could harm future crops.

The following table gives a comparison of the gross content of farmyard manure and seaweed of the *Ascophyllum* variety.

	Farmyard Manure	Fresh Wet Seaweed
	Per 10 Kg	Per 10 Kg
Organic Matter	1.72 Kg	1.81 Kg
Nitrogen	0.05	0.05
Phosphate	0.03	0.01
Potash	0.07	0.12

For a summary of vital trace elements, vitamins and minerals, please refer to the list in the chapter on health and beauty (see p. 97).

Fresh wet seaweed is usually used as soon as it is cut, spread on the land and dug in, whole or chopped. Because nutrient levels vary with season, the best time to gather weed is in the spring with a smaller crop in the autumn. An additional benefit of seaweed-fed land not considered by many agriculturalists in temperate latitudes, but of importance to farmers in the more tropical zones, is the resistance of such soils to leaching of trace elements during torrential rain after long dry spells. The 'stickiness' of the soil also cuts down on topsoil loss due to wind erosion.

In areas where very little soil exists, a 'lazy bed', as the Aran

Islanders call it, can be made from layers of seaweed and sand or sandy soil on bare rock. Primarily used for potato growing, other vegetables take to this form of cultivation, providing subsistence farmers with extra 'land'.

From farm to garden, seaweed is still good news. However, the average garden can be considered more of a closed system compared with the open field and its liberal dosing with raw plants. While rotting, seaweed will use up nitrogen, which can be a problem in a small space, so gardeners prefer to compost their seaweed and use it mixed with other composted materials, peat and sand. The best plants on the compost heap are whole or chopped *Laminaria* and wracks, especially bladderwrack. They provide both humus and plant food, as well as being compost activators. Make a heap of general waste, seaweed and sewage sludge if you can get it. The bacteria will multiply rapidly on the weed, feeding on the sewage and so cause the heap to heat up very fast. Rotting time will depend on the time of year, a good hot heap in summer will be ready for use in about a week.

Seaweed meal (wracks dried and crumbled to a coarse powder) makes a very good organic fertilizer when dug into the soil. If left on the surface it turns into a jelly-like substance, waterproofing the top layer of soil and making life unpleasant for the gardener. Liquid seaweed, whether bought-in or home made, is an ideal foliar feed. It is the nearest thing to 'instant' food that a plant can get, giving a quick boost where needed on anything from oranges to orchids. A friend who ran a local market garden produced super top-quality lettuces all the year round by using liquid seaweed in his mist system throughout the acres of plastic-tunnel greenhouses. His tomatoes were also very popular with local buyers, because they had taste as well as good keeping qualities. If any plant is suspected of having a mineral deficiency, then a quick seaweed spray will put it right. Various fungal diseases respond to the same treatment, such as chocolate spot in broad beans. There is one small problem with using seaweed fertilizer: the potassium present in seaweed locks up the small amount of magnesium present, so you should use another source of magnesium, such as Epsom salts, and do not apply it at the same time as the liquid seaweed.

Many gardening books will show in graphic detail the effects of mineral deficiencies in plants. Lack of boron results in corking on apples, hard and straggly purple coloured cauliflowers, and poor seed setting. Not enough zinc for the plants makes them fail to produce leaves. Manganese deficiency results in pale yellowing plants, and fungus outbreaks. A pathetic little plant with hardly any roots is probably short of copper, and so the list could go on, but the remedies are not so simple.

It is not only the minerals and trace elements that make seaweeds so desirable in the garden. Other substances present can influence cell growth. These are hormones known as auxins, gibberillins and cytokinins. They not only act directly on the growth of the plant, but also seem to act as catalysts to stimulate the plant's own growth hormones. Experiments over many years comparing plants fed on seaweed with those fed on 'synthetics' demonstrated that the latter plants had poorer root development, less leaf area and shorter growth than their seaweed-fed counterparts. More detail of these trials and other information can be found in Stephenson's excellent book *Seaweed in Agriculture and Horticulture*.

Many individual competition growers use seaweed as food for their prize-winning giants. The seed is soaked in liquid seaweed extract for twenty-four hours before planting. Composted seaweed, sand, wood ash, anything else going forms the seed bed. Plants are foliar fed with liquid seaweed as they grow and mulched with composted seaweed at regular intervals. Devout seaweed users will tell you that their flowers have brighter colours and better perfumes. Such claims are difficult to measure scientifically, but thousands of growers can't be wrong.

An experiment was carried out on tomato plants, to compare those grown in plain mediums as against those in the same mediums with added seaweed. A measure of effectiveness was obtained by measuring the amount of manganese present in the leaves of the tomato plants (which is a good indication of a healthy tomato), as is shown in the facing page diagram.

There was a slight drop in effectiveness on sand, but all the other plants showed improved uptake on seaweed treated soils. A side effect which has now been proved by experiment is that seaweed-fed plants show more resistance to adverse weather conditions, especially frost. Again the humble tomato plant was used in these trials, being very frost tender, but weed-fed plants stood up to two consecutive frost days while other plants suffered severe damage. The effectiveness of seaweed in any form incorporated in the soil or spread on top makes it good for rejuvenating old or tired soils that may be covered by a long-lived perennial crop such as fruit trees. Release of bound minerals in

AOTEAROA (N.Z.) TOMATO PLANT EXPERIMENTS

Manganese content of leaves

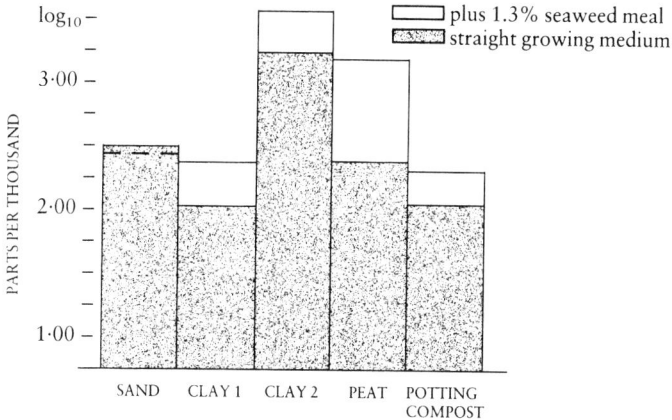

the soil and the addition of nutrients from the decaying weed help to give the fruit trees a boost, soil bacterial activity is increased and subsequent worm movements to incorporate the plants improve the aeration of the soil. In some cases poor or non-producing trees have begun to fruit again. One of the signs of trace element deficiency is that flowers fail to set or produce seed.

Horticulturalists' claims that seaweed-fed plants produced fruits that had longer shelf lives were investigated back in 1964, and shown to be correct. The results of fruit trials at Clemson College (USA) were published in the *Horticultural News* in May 1964 and showed that in parallel runs of fruits from seaweed-fed and unfed plots, after twelve days, thirty-seven fruits had rotted from the untreated plots whereas only eight had gone from the seaweed-fed plants.

Quality is difficult to measure, but we all recognize it when we see it. Seaweed-fed fruits are better 'quality', lawns seem spring-ier underfoot, flowers brighter, carrots more orange coloured, potatoes more abundant and of good shape and so on. It is a qualitative assessment that cannot be matched by mere machine readings; the best way to find out the truth of these statements is to try it for yourself, even if your experiment is limited to two plant pots on the kitchen window sill.

Make your own seaweed fertilizer

Seaweed Meal. The commonest seaweed used in the commercial preparations is *Ascophyllum nodosum.* This is one of the commonest brown weeds and relatively easy to collect. Cut just the tops, leaving the holdfast and stalk attached to the rock. Make this method of cutting a general rule, then stocks of seaweed will be conserved. Add any other weeds handy, especially the browns. Green weed tends to rot too quickly and lose a lot of its goodness, besides making a very unpleasant smell. Spread the cut weed out to dry as quickly as possible. Use the same methods as if you had collected the weed for food. The quicker the drying, the less nutrients are lost. Do not let the rain get on it. Once dry, crumble up into plastic bags. Use some to make alternate layers in the compost heap, together with any seaweed that has not dried, or which you have collected fresh. Mix seaweed powder with compost for potting, or make a good potting medium with sand (from the beach, washed in fresh water), peat and crumbled seaweed. Not too much weed at any one time, or you will have sick seedlings suffering from lack of nitrogen.

Liquid seaweed fertilizer. Two methods using fresh or dried weed. First, take the dried seaweed or fresh cut weed and cut it up fine. Ideally, a liquidizer would be of great use to bring the wet weed down to a pulp. Add a little water to the dried plants if you want to liquidize them. Then put the whole lot into a pressure cooker with enough water to cover and cook for about ten minutes under low or medium pressure. Dilute the cooled sludge by about ten parts of water to one of seaweed. If you are applying the liquid seaweed by spray you must filter the plant porridge before use, through a stocking or a coffee filter paper in a funnel. The filter paper should be pricked all over with a needle before use, or it will clog up before the liquid has a chance to pass through. Put the strained material on to the garden, it is still good food.

The second method is a little more anti-social, involving the rotting down of weeds. Take a five-gallon oil drum, or something similar. In some parts of the country you may be able to get hold of plastic fruit juice tubs and honey barrels from sweet factories. Otherwise use a very strong black plastic sack supported inside a bucket. Use the larger *Laminaria* because they break down very quickly. Put a pile of fresh cut seaweed into your bin. Cover the top with something that will keep the insects out, but allow gases to escape. In two or three weeks, depending on temperature, you will have an excellent plant food, but don't forget to strain it if you want to put the liquid through a spray system. Dried seaweed can be put into the barrel as well, but be sure to add enough water to hydrate the plants; better still, hydrate them before you add them to the bin.

Hints on uses

After all that effort producing a good seaweed fertilizer, how can you use it? As a foliar spray, use it over the entire plant. Many experts believe that the best time to use such methods of feeding is sometime between sunrise and the middle of the morning. Space your sprayings about two weeks apart, so as not to send the plants into shock. Seaweed is such a good food that the best rule is a little and often. If you overdo the applications there will be a reaction from the plants similar to an animal going into shock, hence the term used. It is the same effect as with humans, where we all know that a little of what we fancy does us good, but too much of a good thing can kill us.

Generally, plants fed on seaweed are more resistant to diseases and pests. Carrot fly will not attack rows of carrots dusted with fine seaweed powder. Blackfly and greenfly stay away from well fed broad beans. Champion rhubarb responds well to a seaweed meal before its autumn cover of compost. The hints are as endless as the results claimed for seaweed use in the garden. Whether you gather your own or buy it in ready made, use it on a window box or a thousand-acre farm, the results are the same – marvellous.

Quantities

Fresh seaweed: recommended 25 tonnes per hectare to give approximately 50 kg of nitrogen, 25 kg of phosphate and 50 kg of potassium.

Seaweed meal: 100 to 200 g per square metre, in the autumn.

6

Occurrence, Distribution, Collection and Storage

Light is the chief controlling factor for the depths at which the various seaweeds will occur, but exposure rules the distribution in the tidal zone. Any area will provide a vast range of environmental conditions for weeds but, as explained in Chater 2, there is a general rule of zonation down a rocky shore. There is an overall pattern to the distribution of the various colours of seaweeds down the shore. Nearest the land greens dominate, mid shore to mid deep water browns are found, while reds are confined to the dim light regions. This neat organization has many logical variations, an understanding of which will help you successfully hunt down those special seaweeds.

Green weeds are found everywhere, from brackish pools to the upper levels of the kelp canopy in deep water, but generally they are the first to be exposed as the tide ebbs, draped over rocks or in shallow pools where the light intensity is at a maximum.

Brown weeds display a characteristic zonation down the shore. Above mean high tide level *Pelvetia canaliculata* grows, drying out for most of the time. It is a species that can withstand the rigours of sun, frost and fresh water, but has little tensile strength, so when a few spring tide waves reach it, it breaks up. Even so, the plants can live for four or five years, and only start to reproduce when they are three years old. Lower down the shore comes *Fucus spiralis*, about the neap tide level, then *Ascophyllum nodosum, Fucus vesiculosus* and *Fucus serratus* on the low tide mark. Experiments have shown that spiral wrack (*F. spiralis*) and channel wrack (*P. canaliculata*) are killed if kept submerged in seawater all the time, whereas toothed wrack (*F. serratus*) needs to be underwater for at least six out of every twelve hours. Below low tide mark the large kelps (*Laminaria* species) grow, with only their tops exposed on the lowest spring tide. Brown weeds growing low down the shore cannot stand drying out very much, but they have greater tensile strength than those in the higher zones, so can resist wave action better.

There are always exceptions to these rules, of course. Green seaweeds have been found growing at great depths, and red weeds occur in rock pools on the upper shore, although they still keep to low light areas by sheltering under other weeds and rocky overhangs.

Many of the wracks continue to grow even when torn away from their holdfasts. Out in the Sargasso Sea the drifting weeds have never been rock anchored, and in salt marshes some varieties have no holdfasts, relying on being tangled with other weeds to prevent them being washed away.

Nevertheless, zonation does occur and you will be able to recognize its general layout on any rocky shore. If the shore is very steep then the top levels of weed may be missing, or the zones may be all telescoped together.

Red weeds that are found on the shore also show a zonation effect, although not as distinct as the browns. From near high water mark to about mid shore level the laver (*Porphyra umbilicalis*) can be found. Dulse (*Rhodymenia palmata*) prefers rocks near the low water level, or growing on the stalks of *Laminaria hyperborea*. Under the shelter of the brown weeds and in the lower tidal zone, Irish moss (*Chondrus crispus*) and its culinary substitute *Gigartina stellata* grow in great profusion.

Occurrence and distribution

The quality and quantity of light available not only plays a major part in deciding the zonation of weeds down a beach, but is also one of the factors that affects distribution around the coast. Other governing conditions are salinity and temperature.

Brown weeds seem to prefer the lower temperatures of temperate seas and up into the really cold waters of the Arctic and Antarctic. There are even seaweeds that grow in the gloom under Antarctic ice. Red weeds like it to be somewhat warmer. But wherever they are, the large algae represent a most efficient way of collecting solar energy and converting it to something directly usable by man, with no intervening stage. This is often overlooked in the continuous hunt for ways to 'catch the sun' and use some of the vast amount of sun energy that hits our planet.

Seaweeds like their seawater to be full salinity. If saltiness drops too low, the plants do not grow well, some cannot grow at all, or reproduce. Seawater is to seaweed what soil is to a land plant, and, like land plants, the sea plants grow where conditions suit them best. Underlying rocks and sand affect the holding power of the weed; holdfasts need a firm base on rock or shell and sand is too mobile to give a sound footing. Also there is the effect of sand-blasting by wave-borne sand particles on the seaweeds, and grazing by fish, sea urchins and shellfish. Considering the energy of a storm lashed sea, with waves pounding against the rocks, plus all the other things that go against seaweed survival, it is amazing to see the vast stocks of weed that fringe our shores and survive all that nature can devise for their destruction.

The oarweeds (*Laminaria spp*) grow in cold, deep water, and cannot tolerate more than the briefest of exposure to air at low tide. They only reproduce when the sea temperature is below 16°C, and some cannot tolerate a temperature above 12°C. Deep water with a stony bottom that is relatively sheltered from the worst storm swells is preferred by *Laminaria saccharina*. Dabberlocks (*Alaria esculenta*) is a common seaweed in the northern seas of Britain, but rare in the south. It likes a cold climate and reaches its best development in the winter. Favourite areas for growth are always below low tide mark. Some seaweeds are never uncovered by the tide, and, unless you are a snorkeller or diver, they will rarely be in your collecting bag. Sea sorrel (*Desmarestia*) is one such weed, occasionally washed ashore. It is the one 'nasty' in the seaweed list, containing sulphuric acid esters among other things, so perhaps you should be glad you rarely find it growing.

Locality can control the seasonal viability of seaweeds. In some areas of Scotland, *Gigartina stellata* behaves like an annual plant. In other places it will grow from a cut holdfast for several years. It likes the water to be free of mud, yet well aerated by breaking waves. Near low spring tide level the plant may be found in broad bands. Sewage outfalls are a favoured spot for *G. stellata* because, although it does not like more than the barest minimum exposure to air, direct sunlight and drying winds, it is quite happy in places where the salinity fluctuates as sewage liquid is discharged into the surrounding sea water, and where the bacteria count in the water is high due to the concentration of organic waste. Spore discharge from the plants occurs from September to December, so do not collect weeds in the autumn.

Generally, life histories of seaweeds are complicated and for the enthusiast only, so if you wish to pursue this matter further, see the Bibliography for a way into the literature.

Seaweeds growing on chalk rocks appear to be smaller than

their fellows on other types of rocks, as well as less crowded together. It could be that as the chalk is constantly dissolved by seawater, holdfasts lose their grips when plants reach above a certain size. Wave and current motion will lift the plants away, perhaps with a little flake of chalk still attached to the holdfast. The clean rock exposed underneath will take some time to become 'conditioned' before another sporling settles on it.

Actual distribution of seaweeds depends on where the spores are carried by the currents. If conditions are wrong at the time spores are in an area, then that species of weed will not settle there. Artificial environments may prove to be just what some weeds like most, for example the tropical and sub-tropical species that are now found around the warm water outflows of power stations, and the wracks clinging to the intake pipes where their habit of coming adrift frequently clogs filters.

Search hints
From a single-celled alga 10 μ in diameter to the giant *Macrocystis* at 183 metres (600 feet) seaweeds have a greater range of size than any other group of organisms on earth. Because nutrients are more plentiful in cool waters, the bulk of the seaweeds are to be found in cooler water. Norway has probably the greatest stock of brown seaweeds in Europe, although across the Atlantic, Canadian waters can match Norway for productivity. Britain alone has probably more than 3 million tons of *Laminaria* around its coast. So for both size and quantity you should have no difficulty in finding suitable weeds for whatever your purpose.

The golden rule for collecting is 'know your area'. Get to understand the possible problems of a coastal area. Read the pollution chapter in this book (Chapter 9). Wherever there are people there are usually problems with waste disposal. But do not get too paranoid about one little house up on a cliff top, or even a small settlement. Sewage discharged into the sea is soon diluted and assimilated by the sea life. Look at which direction the tidal flow runs along the coast, then go upstream from the sewage outfall. Try to get about a mile away if you can, and, as you collect your seaweeds, use your nose. Clean weed smells 'weedy', not bad. A quick rinse should be all the precaution you need to take, especially if you are going to cook the plants. Rinse each frond carefully in a rock pool to remove sand grains, then spread it out in a layer one blade deep, with no overlapping. Leave to dry in the sun until it has shrivelled into a 'dead' leaf, taking care to keep evening dew away (see p. 115), or bring the trays indoors at night. If your weed does go mouldy because of the damp, dig it into the tomato patch; tomatoes are always potassium hungry. Once dry, the fronds should keep a long time in a dry jar.

Power stations tend to wash heavy metals out with their cooling waters, shellfish and seaweeds absorb the metals, so treat as with sewage outfalls. The longshore drift of waters around the British coast is to the north and east, so generally go to the south and west of any dubious outfall and let your eyes and nose be your guide.

Tradition has it that laver is not gathered when there is an R in the month. Taste is poor and often covered by fishy or ammonia odours. Many other weeds may suffer from this loss of taste, usually associated with spore production. Summer time is often a low period for nutrient content; spring is best for sugars, vitamins and. young weed to be eaten raw; in late summer and autumn the sugars are often turned to starches for winter food stores, and some plants enter their reproductive phase as the temperature drops, which, as we have seen, adversely affects flavour.

When gathering seaweed, cut the plant above the holdfast. Leave a little stalk and the bottom piece of blade, then the seaweed will have a chance to grow again. Do not overdo the cutting in any one place. In areas where a lot of weed gathering takes place, the recommended time for allowing regrowth is three to five years.

If you find a seaweed that you cannot readily identify, don't pick it unless you particularly want to run it down. Even the most dedicated expert finds it difficult to put a name to some weeds. For food or household purposes, stay with the most common plants (listed in Chapter 2), and the larger ones. This is good conservation practice, and will ensure that there will always be a crop for you or someone else to gather.

One final point: if, in your shoreward ramblings, you turn over stones in rock pools or below the tide mark, please turn them back the right way up afterwards. Thousands of animals and plants live under such stones to get away from the sunlight. Your turning over their shelter means certain death if they cannot move back into the shade very quickly.

7

Everyday Uses: Human Food and Drink

Previous chapters have dealt with the history of seaweeds as foods all around the world, and set out in detail the various constitutents of such weeds. Now we come to the serious matter of eating them.

Survival skills and courses are becoming popular, and one of the easiest places to survive is on the seashore, especially around the British and European coasts that open onto uncluttered ocean and sea. Beside the mountain of man-made rubbish that is washed onto these shores, from which you can easily clothe and shelter yourself, the rocky areas supply a good harvest all the year round to keep you well fed. Eating seaweed and other seafood can become more than just a survival skill if you learn a few recipes and use a bit of imagination. Unfortunately, most of us have little time to spend on the beach so the gathering, preservation and storage of seaweeds for future use is all important.

Preservation and storage
It is not always possible to go out and collect fresh seaweed just when you want to, so, as with other foods, means of storing the vegetable for future use after a good harvesting session are of great importance. Some seaweeds taste better after a certain amount of 'processing', perhaps subtly altering textures to a more palatable form, or just tenderizing. A well dried weed will last many years; all species of seaweed are long lasting if dried. In all of us lurks a squirrel's soul; we need to satisfy the urge that tells us to get ready for the lean months by storing away food. It usually manifests itself in our household by rows of jars filled with wild berry jams, windfall marmalade, pickled eggs, and the freezer bulging with ingredients for future home-made wine. Alongside are boxes filled with dried seaweed, nuts and pulses.

First collect your seaweed
Give it a quick wash in clean seawater to remove unwanted grit

and wildlife, then keep it cool and take it home. The tougher weeds can be heaved into a bucket without any extra water, but keep the more delicate sea lettuces and such like separate in plastic bags. Once home, rinse each weed carefully in fresh water, sorting into types as you go, picking out rough or bad bits. Don't let the plants soak in the water or you will lose valuable nutrients. Most tap water is too acid and can cause the plant cells to burst open. A quick wash is all that should be needed. Some people make up a clean salt solution to wash the seaweeds in, but this can add too much sodium to the system.

Keeping seaweed cool

The obvious way to keep seaweed cool is to put it in the refrigerator. Delicate weeds will keep two or three days, the stronger weeds up to a week. Let your eyes and nose be your guide on this as obviously no two plants will ever be quite the same.

If you want to keep seaweeds fresh while travelling home in hot weather or if you don't have access to a refrigerator, then use a variation on the old-fashioned evaporator once employed to keep milk cool. Originally an evaporator was a coarse earthenware tall pot, soaked in water then stood upside down in a dish of water, the precious milk in a bottle inside. Evaporation caused the internal temperature to drop, so cooling the milk. For seaweeds you could use a large earthenware plant pot (block up the hole in the top) in a large plant saucer. The weeds go inside in a plastic bag with its top open a little. Alternatively, pile seaweeds loosely, then cover with a wet towel, out of the sun but where a breeze can waft by. You can probably think of many variations on this theme, depending on what you have to hand at the time.

The basic techniques for storing seaweeds for home use are freezing and drying; special products such as agars and alginates will be mentioned in the commercial uses chapters.

Freezing

Although freezing is a convenient form of storing and preserving food, freezers are subject to breakdowns, power cuts and gremlins in the system. Nevertheless, for speed and convenience, seaweeds can be frozen. The softer weeds freeze best, with a storage life of about six months before the flavour is lost. Use from the freezer as you would frozen herbs, chopped straight into soups or stews, or defrost with hot water. But remember a certain amount of nutrient loss occurs every time you rinse or blanch the seaweed, unless you use the rinsing water as well. Do not blanch seaweeds before freezing. Tougher weeds like *Laminaria* and kelps do not freeze well. They go strangely rubbery with nasty flavours that no amount of spicing can disguise.

Drying

The most ancient and natural way to preserve food (any food, not just seaweed) is drying. Nature dries things out beautifully on the beach on a hot summer's day, but unfortunately such weather is not always around when we need it. To sun-dry seaweed, just rinse, bunch the stronger, branchier types into small bundles and hang from the branches of a shady tree, in the breeze. Large kelp fronds should be hung up singly. Sea lettuce and laver bread types can be spread out on flat stones, over wooden frames, Japanese style, or just heaped into small mounds, frequently fluffed up and turned so no damp bits remain hidden away. Bring all the seaweeds in each evening before the dew falls. Carragheen needs to be put in the direct sunlight before it bleaches out to a tawny beige colour: it takes several days.

For less gentle climes, build a solar dryer. It can be used to dry anything from seaweed to apricots, and protects the food from contamination as well. A good basic design can be found in John Seymour's book *The Complete Book of Self-Sufficiency*, with trays modified to take seaweed. Make trays from wooden frames onto which thin fine mesh has been stapled or sewn. The mesh can be old stockings opened out, net curtains, or even cotton thread wound back and forth between nails on the frame. Spread weeds loosely, no two bits overlapping, so that the air can circulate freely. Cut up very 'twiggy' bits so that they can make best use of the space available. Split thicker stalks into strips. Lay flat lettuce-type fronds out onto mesh, uncurling folds and smoothing down as much as possible. Test for dryness frequently, removing those that are ready as soon as you can. The solar dryer described has stones in it which act as night-storage heaters, thus keeping off the dew, continuing the drying process after the sun goes down, and saving you the job of unloading all those trays every evening.

Drying seaweeds in the house, if you have a kitchen range, is very easy: put a rack of some sort over the stove, perhaps one of those old plate warmers with a cloth on it, and spread the plants out in sheets or hanging in bunches from the hooks. Beware of cooking the weed before it is dry, don't let it get hot, just warm. Even better are the old fashioned drying rails that were once so common in large kitchens, and which are still used in some hotels for keeping tea-towels aired. A series of rails, joined by cross-pieces, all hung on ropes and pulleys to be raised to the ceiling where the warm air is, or lowered for loading.

If you just want to keep a small amount of seaweed, or are just starting to experiment, trays in an oven after the baking is finished will suffice, but again, be sure the oven temperature is not too high.

Seaweed needs to be dried as fast as possible, preferably in one go. In *The Sea Vegetable Book*, Judith Madlener recommends about twelve hours to crisp dry, in order to maintain the value of the vitamin A in the plants.

Microwave cookers can be used to dry seaweeds. In experiments with a low-powered microwave cooker (only 300 watts) sea lettuce took three minutes, and bladderwrack seven minutes – and frightened the dog. The bladder gasses expanded and exploded just as the dog came into the kitchen sensing 'food'. She never trusted that microwave cooker again, and wasn't at all impressed by the scientific aspect of the experiment. Generally, thick stalks of the tougher weeds such as kelps seemed to stay tough, even when rehydrated. Their composition was somehow changed, but the blade areas were all right and flavour seemed unaffected. In the sea lettuce type of weed the flavour was, if anything, enhanced.

Hot air drying, other than by solar power, is possible at home, if you don't mind spending money. Weed is dried commercially this way, by blowers and electrically heated air. A home version can be made with electric light bulbs in the bottom of a box fitted with trays. Protect the bulbs from accidental water splash by putting a solid tray between bulb and weeds, which will also keep most of the light off the weeds at the same time. Using electric heaters to dry the crop could make it cheaper to go out and buy your seaweed from the local health shop, unless you generate your own power.

Improving the flavour

When you have just picked a good pile of seaweed and are preparing it for drying or freezing, then is the time to indulge in a little 'improvement' of some of the flavours of the weeds. Brown seaweeds respond to a quick blanch in boiling water, and, although a little of the nutrient is lost, there is a marked improvement in flavour, as well as making it quicker to cook once rehydrated. Microwave cooking/drying may have the same effect without the loss of any nutrients. It is an area of interesting possibilities yet to be explored.

Smoking over hardwood (not pine or fir) chippings can give the most glorious flavour to a brown seaweed. If you use softwood sawdust the food acquires a resin taint, a bit like turpentine.

Storing

Seaweeds should be stored in airtight containers in the cool and in the dark. As with all dried foods, a little damp can destroy them, but regular checking of the stock will deal with problems before they get too serious. Throw away mouldy bits, but don't mistake dried salt patches for mildew. Dry off any that seem damp in a slow oven when you have finished baking, or put a tray over the stove, in a sunny window, airing cupboard, or waft quickly across the flames of the fire until crisp and dry again. Cats seem to think that the very best place next to heaven is a tray of warm seaweed on a sunny window sill.

Avoid aluminium containers; stainless steel is all right, glass is best, but plastic bags enable you to peg the tightly closed and sealed packages with their air sucked out, onto a line in cupboard or pantry, leaving valuable shelf space free.

Whatever your storage method, use the smallest containers possible, each one perhaps holding only a single helping. This ensures minimum air contact with the dried weed until it is actually used. If you can suck the air out before sealing, as with plastic bags, that is good, otherwise pack as much into the container as you can, so leaving very little room for damp air to creep in. Do your packing and sealing in a warm room where the air is dry, not when it is pouring with rain outside, the kitchen is full of damp washing, and dinner is boiling merrily on the cooker. You would be surprised at the amount of damp in that sort of atmosphere just waiting to get at your seaweed and make it wilt. After all, that is why the weather forecasters have their precious bits of seaweed to tell them when it is going to rain.

Now you are ready to try some of the recipes, but a word of advice. Try small quantities at first. As with all foods, our taste buds need time to adjust. Don't dismiss a weed out of hand after the first mouthful; you really should sample at least three meals before you finally give the thumbs down. Seaweed tastes unusual but rarely unpleasant. Remember, all the foods we take for granted now and enjoy must have been tried for the first time by someone. Imagine a mouthful of strong green cabbage as your first introduction to the brassica family; not a good way to get acquainted, yet now we all accept cabbage, cauliflowers, Brussel sprouts, turnips and mustard as everyday food. It was a brave soul who tasted the first potato, knowing probably that the leaves of that same plant were poisonous (it is related to deadly nightshade). So take heart and give seaweed a try.

8

Recipes

The recipes in this section are mainly vegetarian, but some include fish or meat where the dishes are Far-Eastern favourites. However, all the recipes are easily adapted to the vegetarian taste; meat can often just be omitted without any substitute flavour being required. This is especially true of soups. Valuable lessons can be learned from the few meat and fish recipes included, especially those from the Chinese influenced areas. They illustrate the supreme economy of food use, where truly nothing is wasted. This is also in the tradition of Cordon Bleu, where only the best was used, but not even the tiniest crumb was left uneaten. To waste food goes against all the principles of conservation.

Chinese cooking is a model to be copied wherever possible. Essentially, it seeks to cook vegetables so that the ingredients and flavours remain individual and recognizable when combined in one dish. Therefore preparation often takes a lot longer than cooking and the minimum amount of cooking is used to bring out the best in the ingredients. Sea vegetables are so succulent and tender that this quick cook method is ideal for nearly all the species. Flavour is not spoiled, and nutrients are conserved.

In all these recipes, though one weed is specified, any other weed of the same type can be substituted where the flavour is likely to be the same; for example, sea lettuce and *Enteromorpha* can be interchanged. *Laminaria saccharina*, however, is in a class of its own.

Seaweed recipes are arranged in their groups: first the fresh green weeds, then fresh browns and reds; then the imported dried weeds available in health stores in this country, recipes for which are available on the packages themselves. Finally a section dealing with the ubiquitous agar.

For your information, the recipes have, wherever possible, been calorie counted, and fibre and carbohydrate units calculated, for the whole meal. All dishes are for two people.

WEIGHTS AND MEASURES

250 ml = 8 fluid ounces = 1 cup

15 ml = 1 tablespoon = ½ oz

5 ml = 1 teaspoon

1 USA tablespoon = 3 British teaspoons

45 g = 16 oz

2.8 g = 1 oz

3.8 l = 8 pints

0.24 l = ½ pint

30.5 cm = 12 inches

1 bar of Kanten (solid agar) = 0.7 g = ¼ oz

1 bar is approximately 2.5 tablespoons

1 bar will set 5 cups of liquid

½ cup Irish moss will set 4 cups of liquid

1 level teaspoon of agar will set 0.24 litres (½ pint) of water

RECIPES

CHLOROPHYCEAE (Green seaweeds)

Chaetomorpha (Hair greens)
Favourite as a condiment. Quickly rinse the weed in fresh water, then dry thoroughly. Powder, and keep in an airtight container.

Salad dressing (256 calories; 0 fibre; 0 carbohydrate)
2 tablespoons sesame oil
juice of half a lemon/lime
shake of *Chaetomorpha* powder
pinch of brown sugar

Blend well in a liquidizer, and dress a green salad with the mix. Salads improve if they are put in a refrigerator for half an hour before serving. This gives the flavours of the dressing time to blend and mellow with the vegetable.

A quick fillip for a commercial mayonnaise is to mix in *Chaetomorpha* powder and a little garlic powder to make a beautiful green dressing. Sprinkle green powder over prawn cocktails, or flavour fish pie.

Salad
Chaetomorpha greens are tasty chopped into a salad, but do be sure that they are clean.

Enteromorpha species
Gather fresh weed and rinse quickly in fresh water. Waft the fronds over a flame so that they toast but do not scorch. This brings out the best flavour.

Fish Grill (in 3g: 65 calories; 0 fibre; 0 carbohydrate)
2 oily fish such as mackerel, really fresh
dried and toasted *Enteromorpha*, about 2 tablespoons
twist of black pepper

Gut and split the fish down their length. Put onto a grill or barbecue skin side up, and cook quickly until the skin bubbles. Turn the fish over, sprinkle with *Enteromorpha* and a little black pepper, close up the two halves, and finish cooking. Serve on a bed of green salad in which fresh *Enteromorpha* is included.

Use all species in soups and as salad vegetables on their own or included in mixed dishes. *Enteromorpha* is more versatile than garden lettuce.

Codium tomentosum
A fine flavoured plant that needs to be washed carefully in lukewarm water to remove debris from its velvety surface. To preserve it, chop the plant very fine, then dry rapidly or it will decompose. The Koreans use *Codium* as we would use garlic.

Codium salad (130 calories; 1 fibre; ½ carbohydrate)
½ teaspoon fresh *Codium* plant
the green half of a Chinese leaf, or a whole lettuce
1 medium onion
juice of half a lemon
2 teaspoons pineapple juice (unsweetened)
3 teaspoons sesame or olive oil

Chop *Codium* plant, Chinese leaf or lettuce, and onion as fine as possible. Mix oil, lemon and pineapple juice well, then pour over the salad. Toss till well coated and chill slightly before serving. A few thin slices of tomato for garnish give a pleasing colour contrast.

Korean tea
Take washed *Codium* plant and dry it as fast as possible. When thoroughly dry, reduce to a powder and store in an airtight tin.

To make the tea, warm a china pot. Put in 1 teaspoonful per person. Add water that has just gone off the boil. Leave to brew for two minutes. Serve straight or with a dash of lemon. The tea is spicy and sharp.

Ulva lactuca
Gather fresh, bright green leaves, rinse quickly in fresh water, then use either fresh or dried. This is the famed green nori of Japan, also known as green laver.

Ulva salad (in 30 ml, 2 tablespoons, 2.8 g: 15 calories; 1 fibre; ½ carbohydrate)
Take fresh leaves, chopped fine. Dress with a little fresh lemon juice. Serve as a dish in its own right.

Cantonese style seafood soup (200 calories; 3 fibre; 4 carbohydrate)
Vegetarian variation: use a stock of barley miso for extra flavour.
2 large leeks
2 medium onions
generous dash of soy sauce
½ glass light sherry
a little oil
small handful of fresh sea lettuce
empty shells of crab, lobster, tiny whole crabs, shellfish shells.

Slice the leeks and onions very fine. Stir-fry in the oil for 2 minutes (the less oil used, the fewer the calories). Stir in chopped seaweed. Boil all the shells in enough water to cover, for 20 minutes. Strain. Add the liquid to the stir-fried vegetables. Season to taste and serve hot.

Shrimp soup (182 calories; 3 fibre; 4 carbohydrate)
Vegetarian variation: just leave out the shrimps, and reduce the calories by 35.

150 g (5 oz) fresh shrimps
50 g (2 oz) bamboo shoots
50 g (2 oz) small hard cucumbers
50 g (2 oz) watercress
10 g (1/3 oz) sweet red pepper
10 g (1/3 oz) *Ulva* or other green seaweed
10 g (1/3 oz) red seaweed (any type)
1 slice fresh root ginger
½ cup white wine
750 ml (1.1 pints) clear stock (shrimp shell, chicken or vegetable)
a little oil

Slice all the vegetables very thin, and stir-fry for 2 minutes. Boil the shrimps, ginger and bamboo shoots for 2 minutes. Put in the vegetables. Simmer for a further minute, skimming off any scum. Remove from the heat, add the wine and serve.

Bryopsis plumosa
Wash the weed quickly in fresh water, then dry rapidly for use as a condiment

Sentinel soup (440 calories; 15 fibre; 70 carbohydrate)
15 g (½ oz) split red lentils
2 medium onions
a little oil
1 level teaspoon *Bryopsis* powder

Cook the lentils in enough water to cover twice, until they are reduced to a cream. Fry the thinly sliced onions in the oil till soft. Mix onions and *Bryopsis* powder into the lentil soup, reheat gently and serve with fingers of buttered toast.

Mushrooms de mer (110 calories; 1½ fibre; 1½ carbohydrate)
4 'wide awake' mushrooms (large, open ones)
small handful of fresh *Bryopsis*
4 large tomatoes
enough grated cheese to sprinkle over

Wipe over the caps of the mushrooms with a damp cloth. Remove stalks. Chop the stalks and the tomatoes finely. Arrange the mushrooms in a casserole dish, curl the *Bryopsis* round on each cap, letting the feathery fronds drape over the edges a little. Pile tomato and stalk mixture on top. Cover and microwave (7 minutes on 300 watts). Then add the sprinkle of cheese. Finish off in microwave cooker, or pop under the grill. Serve with a green salad.

PHAEOPHYCEAE (Brown seaweeds)

Alaria esculentia
Eat the sporophylls fresh, for a delicious nutty flavoured snack. Cut out the midrib and eat that fresh. Dry the blade to remove 'off' flavours.

Nutty salad (40 calories; 3 fibre; 2½ carbohydrate)
Large handful of *Alaria* sporophylls, fresh
1 teaspoon of toasted sesame seeds
¼ white cabbage
1 Chinese white radish

Shred the cabbage very finely. Grate the radish and mix into the cabbage. Chop up the sporophylls coarsely, and toss in the toasted sesame seeds. Pile the *Alaria* and sesame mixture onto the salad bed. Serve with a variety of dressings, such as lemon juice, mayonnaise, and light soy sauce.

Alaria ribs in cheese sauce (130 calories; 2 fibre; 2½ carbohydrate)
10 *Alaria* midribs, fresh
1 medium tin vegetable soup
50 g (2 oz) strong (blue is best) cheese
a dash of white wine

Cut the ribs into finger-length pieces and arrange in a casserole dish. Heat the soup, stir in cheese over a very slow heat until it has completely melted. Remove from the heat and add a dash of wine if liked. Pour over the *Alaria* ribs. Cook for 5 minutes in a microwave cooker or until the mixture just bubbles. Serve hot with creamed potatoes.

Comforters (Per piece: calories 90, fibre ½, carbohydrate 22)
Finger-length pieces of *Alaria* midrib, dipped into honey, rolled in toasted oats and sesame seeds.

Alaria stew (600 calories; 30 fibre; 90 carbohydrate)
200 g (6 oz) dried *Alaria* blade
200 g (6 oz) carrots
200 g (6 oz) parsnips
200 g (6 oz) swede
200 g (6 oz) potato
any leftover vegetables from other meals, even salad stuffs
1 tablespoon barley miso (or 2 teaspoons Marmite, Barmene or similar)
2 tablespoons concentrated tomato purée

Soak *Alaria* in enough cold water to cover – overnight is the usual practice. Chop or mince all the other vegetables. Drain water from the soaked *Alaria*, and use it to mix in the tomato purée and miso. Add warm water until there is enough to ensure two generous helpings of stew. Bring gently to a simmer, adding first the *Alaria*, then the other vegetables. Cook until all are tender; this will depend on how finely chopped the vegetables were to start with. Pressure-cooking reduces cooking time, and conserves nutrients by reducing their opportunity to oxidize.

Extras
Rehydrated dried midrib can be fried until crisp, and used as dip in cheese sauce.

Ascophyllum nodosum

Knotty rolls (1,000 calories; 100 fibre; 200 carbohydrate)
six slices of brown bread
margarine
large handful of fresh young *Ascophyllum* plants
2 medium eggs
a little brown flour or breadcrumbs
oil for frying
12 wooden cocktail sticks

Spread margarine on one side of each slice of bread. Steam the *Ascophyllum* plants until tender. Make each slice of bread into a roll filled with *Ascophyllum*, held by two cocktail sticks threaded through. Beat up the eggs. Dip each roll into beaten egg, roll in flour or breadcrumbs and fry rapidly in shallow oil until crisp. Nice for parties, barbecues, etc.

Other uses
Stir-fry steamed *Ascophyllum* with other vegetables and use to stuff pancakes. *Ascophyllum* makes a lovely vegetable on its own, goes well with dark soy sauce, cheese sauce and tomatoes. Flavour is complementary to peanut butter as a filling in toasted sandwiches.

Chorda filum
Wash well, especially if using cast weed. You can scrape off the 'slime' if you feel you won't like it.

Salad on a string (170 calories; 4 fibre; 3 carbohydrate)
six fresh tomatoes
a little green pepper
a dash of garlic powder
touch of olive oil
juice of half a lemon
generous handful of fresh *Chorda*

Chop up tomatoes and *Chorda*. Reduce green pepper to a fine mince. Mix pepper, lemon juice, oil and garlic powder. Put the *Chorda* on a plate, arranged in a layer, like spaghetti. Pile the tomatoes on top, and dress with the mixture. Do not be tempted to over mix. Keep it shiny and sharp.

A special event (150 calories; 5 fibre; 9 carbohydrate)
handful of long *Chorda*, steamed tender
pint of made-up agar jelly (see p. 00)
1 tub cottage cheese (large)
mixture of cooked peas, sweet corn, and tomatoes

In a glass dish arrange the *Chorda* filaments into pretty patterns such as plaits, bows, names, fishes, etc. Spoon a little liquid jelly over the designs to just cover and hold them until set. Mix the cottage cheese with more agar jelly, and arrange over the layer of *Chorda*. Leave to set again. Fill mould up with remaining jelly and vegetables, and put to set firm. Dip mould quickly in warm water and turn out onto a bed of lettuce.

Ectocarpus species
All very strong flavoured, so used as condiment. Dry the weed well, grind to a powder and store in a dark jar. Do not be over-generous when using *Ectocarpus*, or it will drown out every other flavour.

Ectosoup (175 calories; 2½ fibre; 18 carbohydrate)
3 large onions
1 large potato
1 small clove garlic
pinch of *Ectocarpus* powder
500 ml (1 pint) stock
little butter or margarine

Cook the potato or use left-overs from a previous meal, and mash. Chop garlic and onions finely and sauté till soft in the butter or margarine. Mix in potato and *Ectocarpus* powder. Beat in the stock. Bring to simmer slowly, keeping the soup pale cream in colour. A little milk can be added if liked. Serve hot with croûtons.

Ectocarpus powder can be used to make a fish soup without the fish, or as a flavour enhancer on oily fish such as mackerel. Use a mixture of ground ginger, garlic and *Ectocarpus* powder to add spike to vegetable burgers for a barbecue party.

Fucus vesiculosus
The famous bladderwrack, good as both food and medicine. Although it seems a tough plant, it needs only a little cooking. The plant itself is not eaten, all its goodness is stewed out, the resultant stock used and the empty plant fibres consigned to the compost heap.

Potato pot soup (100 calories; 1 fibre; 24 carbohydrate)
2 large potatoes
1 whole fresh or dried bladderwrack plant
enough water to cover potatoes in a pot

Chop potatoes and put to simmer in a pot. Add the whole wrack. Cook until the potatoes fall apart. Remove the weed and discard. Mash down any remaining lumps. Serve hot and smooth.

Pasta sauce
4 tablespoons tomato purée
2 tablespoons barley miso
pinch garlic powder
2 tablespoons soya protein granules
1 whole fresh or dried bladderwrack plant

Put the wrack into a stainless steel pot and simmer in enough water to cover for 30 minutes. Discard the plant. Into the liquor mix the tomato purée and miso, adding more hot water if required to make a runny stock. Put in the garlic powder and soya protein granules. Stir well and leave to stand so that the soya soaks up much of the juice. Add liquid to make a thick but runny mixture, and simmer for about 5 minutes. Serve on a bed of wholewheat spaghetti, top with cheese if desired.

Other uses
The liquor extracted from cooked bladderwrack is good in any stews or soups. Dried lateral branches are used to make tea: 1 teaspoon of dried weed to 1 cup of boiling water, sweetened with honey if required. Clears arteries and improves circulation.

Scytosiphon lomentaria

Winter reviver
50 g (2 oz) dried *Scytosiphon*
1 tablespoon barley miso or Marmite

Soak the weed for 15 minutes in a little cold water, then drain. Use the water as a face wash, especially if skin is chapped. Put the miso into one pint (575 ml) of boiling water, add the weed and simmer for two minutes. Drink while hot. Be careful if using Marmite, as it is very salty.

Stew in the green
generous handful of dried *Scytosiphon*
50 g (2 oz) frozen peas
50 g (2 oz) frozen string beans, or similar
pinch of dried spearmint
dash of garlic powder
50 g (2 oz) egg noodles

In a large cooking pot (not aluminium) put all the ingredients, with enough water to cover. Bring gently to the simmer, and add any water to keep the vegetables just covered. Cook until noodles are soft. Serve hot. This dish takes advantage of the delicious bean-like flavour of the seaweed. It acts as both vegetable and seasoning at the same time.

Other uses
Scytosiphon has such a nice taste that it can be used as a main vegetable in its own right, if you can gather enough.

Laminaria digitata
One of the kelps. The tastiest way to treat kelp is to smoke it cool over an oak fire, as you would a good kipper. When dry, press into cakes, then slice thinly to form threads. Rehydrate before use unless putting into a liquid dish such as soup.

Smoky soup (90 calories; 4½ fibre; 15 carbohydrate)
50 g (2 oz) split peas
20 g (1 oz) smoked *Laminaria* slivers
1 large onion
Soak the split peas overnight. Chop up the onion coarsely. Put peas, onion and *Laminaria* slivers into a pressure-cooker basket. Cover with water and cook under low pressure, or in a saucepan on a medium heat until the peas disintegrate. Add more water to make a smooth soup. Traditionally, bits of bacon were used as the flavouring in both pea and lentil soup, but you will find that the smoked *Laminaria* has a finer flavour.

Salt substitute
Use up any left-over bits of *Laminaria*, dried in a slow oven until crisp, but not scorched (that spoils the flavour). Grind to a fine powder and use instead of salt. Store in small quantities in very small containers, so that the whole does not get damp and waste your effort.

Laminaria longicruris
Oarweed or weatherman's weed; can be eaten fresh, or preserved by drying. Especially good if cool smoked over oak chippings.

Boatman's parcels
50 g (2 oz) brown rice
50 g (2 oz) soya protein mince
1 medium onion
2 teaspoons barley miso
2 teaspoons tomato purée
1 or 2 long fronds of oarweed

Cook the rice. Soak soya protein in miso and tomato purée made up with a little hot water. Sauté finely chopped onion till soft, then stir in the soya and rice, mixing well. Cut oarweed into squares about 10 cm (4 inches) wide. Put a generous heap of mixture in the centre of each square, then fold over corners to make a parcel. Pack each parcel, folded side down, into a steamer or pressure-cooker basket. Steam for 30 minutes or pressure-cook for 10 minutes on low pressure. The parcels can also be microwave cooked in a closed glass casserole for 5 minutes, with a sprinkle of stock added to keep the weed moist. Serve with soy sauce, or sprinkle with cheese and grill. Spoon any left-over filling around the parcels or freeze for future use. The seaweed parcels freeze well, before or after cooking.

Other uses
Take squares of fresh weed, dip in egg batter and quick deep fry.

Laminaria saccharina
Laminaria contain a natural taste booster similar to sodium glutamate, and in addition *L. saccharina* has mannitol, a sugar, which gives it its popular name of sugar wrack. The sugar gives this weed a special sweet flavour all of its own.

Sea nest
large frond of sugar wrack
50 g (2 oz) brown rice
1 large parsnip

Cook the brown rice. Cut sugar wrack into long thin 'laces', and simmer till tender. Slice parsnip very thin, then quick deep fry to make crisps – do not let them brown too much or you lose the sweetness. Arrange the sugar wrack on a bed of rice to make a nest, then fill with parsnip chips. Serve with a sea lettuce and lemon salad.

Pickled sugar
sugar wrack frond
1 medium onion
1 Chinese white radish
1 thin slice of ginger
dash of turmeric
vinegar

Roll the sugar wrack into a tight roll, then slice thinly. Slice the onion paper thin, do the same with the radish, plus any odds and ends of vegetable such as the hard white stalk in the middle of a cabbage. Sauté all in a little oil, stir in the turmeric and ginger, then pack into a jar. Cover with a vinegar of your choice, tapping the jar to remove trapped air. Cider vinegar is nice, but avoid the stronger flavoured herb vinegars. Leave in a dark place for about 2 weeks.

Other uses
Eat the stalks of fresh weed or deep fry. Pieces of frond, either fresh or smoke dried and rehydrated, are fried till crisp, served as an appetizer before a meal, or with party dips.

Pelvetia canaliculata
A strong flavoured, fat, rich weed that grows at the top of the shore. Dried and used as a condiment.

Banjaxed onions
2 large Spanish onions
1 cooking apple
1 small glass whisky
1 tablespoonful raisins

Soak the raisins and chopped apple in whisky for 2 hours. Peel the onions, leaving the base held together by the root area. Hollow out from the top, leaving two or three layers of onion around the outside. Fill with the fruit and pour any remaining whisky over the onion. Sprinkle with *Pelvetia* seasoning. Cover close and either bake or microwave until soft. This mixture also enhances potatoes in their jackets.

Petalonia fasciata

Bean balm pot
250 g (9 oz) haricot beans
150 g (5 oz) dried *Petalonia*
2 medium onions
a little butter
1 litre (1¾ pints) stock
3 tablespoons tomato purée

Soak the haricot beans overnight in plenty of water, then cook until soft, at least 30 minutes under pressure, 2 hours' simmering in a saucepan. Pressure-cooking is ideal for this process. Soak the *Petalonia* until it relaxes (only a few minutes), then chop into small pieces. Slice the onion and sauté in butter till transparent, stir in *Petalonia* and cook for 2 minutes, but do not brown. Add beans, stock and tomato purée. Either simmer slowly until much of the liquid has been absorbed, or cook in a casserole for at least an hour.

Other uses
Petalonia is ideal in stews and casseroles. It is a substantial vegetable even in small quantities. Fresh leaves can be blanched by pouring boiling water over them in a colander, then used in any cooked dish.

RHODOPHYCEAE (Red seaweeds)

Ahnfeltia plicata
Although many of the red weeds are agar source plants, they do have other food uses. Therefore each weed will be dealt with as an individual vegetable in this section and agar has a larger section of its own later in the chapter (see p. 00).

Pit chicken (42 calories per 2.5 g)
1 fresh chicken
a good supply of fresh *Ahnfeltia*
melted butter

Clean, wash and dry the chicken. Put in any stuffing desired. Wrap the bird in *Ahnfeltia* until it is completely covered. Put in a large casserole or chicken brick. Pour over melted butter until the *Ahnfeltia* is well oiled. Bake slowly until the chicken is cooked, at least 3 hours.

Vegetarian pit roast
50 g (2 oz) assorted nuts (especially hazelnuts)
2 eggs
50 g (2 oz) brown breadcrumbs
50 g (2 oz) assorted diced cooked vegetables

Grind the nuts to a fine mince, mix in the vegetables and

breadcrumbs. Beat the eggs, stir well into the mixture to make a thick roll. Wrap in *Ahnfeltia* and butter or margarine as for the chicken, cook slowly, well covered to prevent drying out.

Other uses
Fresh plant is often served as nibbles. Very nice in salad. A good vegetable to throw into a casserole.

Bonemaisonia asparagoides

Sea balls
handful of fresh *Bonemaisonia*
25 g (1 oz) cornflakes
1 egg
100 g (4 oz) flour
a little milk, butter and salt

Wash the fresh *Bonemaisonia*, chop small, then pound to a pulp with the salt and a little butter. Mix in crushed cornflakes. Make into balls about the size of a walnut. Beat the egg, add flour and a little milk to make a thick batter. Dip each ball in batter and deep fry till crisp. Serve with sea vegetable salad.

Other uses
Chop plant into salads, or simmer in a little butter then serve with jacket potatoes.

Chondrus crispus
Irish moss or carragheen. The plant has a stronger 'seaweedy' smell than most others, because of its high sulphur content. Taste varies with the length of time it is allowed to soak before cooking; ten minutes is enough for use in strong savouries, but a change of water and another ten to fifteen minutes' soak is needed if the moss is to be used in sweets or more delicately flavoured dishes. Irish moss is interchangeable with agar in most recipes, but you will need to use twice the weight. Where a recipe calls for any food containing wine vinegar or oxalic acid such as in chocolate, spinach or rhubarb, then use Irish moss instead of agar.

If you gather your own moss, the late summer plant is highest in carragheenan, and needs to be put to dry and bleach in the sun until creamy white.

133

Seashore soup (50 calories; 1 fibre; 2 carbohydrate)
25 g (1 oz) chopped, sunbleached and dried Irish moss
1 litre (1¾ pints) water
250 ml (½ pint) white wine
pinch of tarragon
1 tablespoon barley miso
100 g (4 oz) assorted bits of other seaweeds

Soak the moss in cold water for 10 minutes, rinse thoroughly. Put into a pan (stainless steel is best) with the water and tarragon, bring to the boil, then simmer for about 20 minutes. Add the miso, mix well. Remove from the heat and stir in the wine. Strain through a jelly cloth. Serve the soup with fingers of garlic toast.

Seaweed pudding
small handful of sunbleached and dried Irish moss
0.75 litre (1 1/3 pints) whole milk
1 vanilla pod
1 egg, and the white of an egg
1 tablespoon caster sugar

Soak the moss in cold water for 10 minutes, then cook gently in the milk (with vanilla pod) for 10 minutes. Whisk together the egg and sugar in a large bowl. Strain the moss and milk through a sieve into the egg mix, pushing through the thicker parts of the jelly with a wooden spoon. Whisk well. Also whisk the egg white to a stiff peak, fold into the moss mixture and put somewhere cool to set. Serve with Irish whisky sauce.

Nettle beer
450 g (1 lb) fresh young nettle tops
225 g (8 oz) sugar
10 litres (17 pints) water
1 tablespoon malt extract (for a light beer)
4 sprigs sunbleached and dried Irish moss
25 g (2 oz) dried hops
15 g (½ oz) dried yeast

Boil nettles and hops in a loose net bag in the water, for 20 minutes. Remove from heat. Add sugar and malt extract. Leave to cool. Remove bag of hops and nettles. Add Irish moss and yeast. Cover and leave to stand for about 18 hours. Siphon into bottles through a coarse sieve and cork. Put the bottles somewhere cool, and where any explosion will not cause too much damage. Relieve excess pressure by just easing the corks, or use special pressure-relief plugs. Drink after three days. It is as alcoholic as home-made ginger beer, and tastes very refreshing, rather like a mild ale.

Gelidium species
These plants are usually the source plants for agar (see p. 141). Fresh plants tend to have a mousey taste, which has to be removed by sun drying, or bleaching out the red colour. They are incorporated into dishes as jellying agents, the degree of *Gelidium* flavour remaining being dependent on how much processing of the weed has been carried out.

Gelidium soup
1 bleached gelidium plant
2 medium onions
2 large carrots
2 sticks celery
0.5 litre (¼ pint) stock
a little butter or oil

Slice the vegetable thinly and sauté in the butter or oil. Pour in the stock. Put the whole *Gelidium* plant into the mixture and bring gently to the boil. Simmer for 20 minutes. Throw away the *Gelidium* plant. Serve the soup hot with fingers of toast and peanut butter.

Gigartina stellata

Vegetable spread
1 medium hard cucumber
1 large egg
1 tablespoon oil
2 teaspoons (mean) *Gigartina*
1 medium onion
2 medium carrots
pinch of paprika
1 teaspoon lemon juice
500 ml (1 pint) water

Grate the onion, carrots and cucumber into a large bowl. Add

135

paprika, lemon juice and mix well. Boil the *Gigartina* in the water for 20 minutes, then set aside to cool. As it begins to gel, beat in the egg and oil. Pour over the vegetable mixture and stir well. Put in a covered bowl in a cool place to set, and allow flavours to blend. Use as a filling for sandwiches, cocktail snacks, etc.

Gracillaria verrucosa
The British plant is more tender than the American species. In some recipe books you will find this species referred to as 'China grass'.

Coconut sweet
50 g (2 oz) sun dried *Gracillaria*
100 g (4 oz) desiccated coconut
0.5 litre (¼ pint) coconut milk
squeeze of lemon juice
8 tablespoons of rum
1 tablespoon brown sugar if desired

Soak the *Gracillaria* in coconut milk, bring to the boil slowly, then simmer for 20 minutes. The milk will thicken a little. Mix coconut, rum, lemon and sugar if used. Stir in the *Gracillaria* milk. Put in a cool place to set. Serve as it is, or scoop out the centre and fill with exotic fruits, ice cream or flavoured ices.

Other uses
Blanch the fresh weed by pouring boiling water over it in a colander. Chop fine and add to rice salad, with a dash of soy sauce and slivers of ginger.

Nemalion helminthoides
Common name for this plant in Italy is 'Turkish spaghetti'.

Sea foam soup
2 cups fresh *Nemalion*
4 egg yolks
2 tablespoons rice wine
2 tablespoons light soy sauce
4 tablespoons light miso
4 cups boiling water

Blanch the fresh weed, then chop finely. Whisk egg yolks to a thick cream. Stir the miso into the boiling water and allow to cool a little. Whisk wine and soy sauce into the egg yolks, add miso liquid. Pour the whole mixture over the *Nemalion*, either in individual cups, or serve in a large tureen.

Other uses
Chopped, blanched weed is nice in salads. Rehydrated weed is also good.

Dulse (*Rhodymenia palmata*)
A widely used, most popular seaweed; *Rhodymenia palmata* is a favourite among many coastal people in the UK, around North America, and Japan. It has the highest concentration of iron of any food source, as well as potassium and magnesium, so it is no surprise that as well as being eaten for food, it has extensive medical uses.

This versatile and tasty vegetable is usually gathered by hand during May to August, sun dried and packaged without processing, so you need to keep a weather eye open for bits of shell or stone when the fronds are rehydrated.

Picking your own dulse is highly recommended. Once you and your family develop a real taste for it, then every trip to the seashore will be economically sound as well as a great deal of fun. A bucket of weed will probably feed a family for a week or two.

Celtic drummer (A Cornish treat)
A summer substitute for bacon and egg pie.

½ cup dried dulse
3 eggs
145 ml (¼ pint) milk (goat's milk if possible)
pastry case in a flan or pie tin (7-inch diameter)
1 large leek or onion
sprinkle of thyme
a bit of butter

Freshen dulse in cold water then drain and pat dry in a clean tea-towel. Sauté onion, thinly sliced, in the butter until transparent, then stir in dulse and fry for about five minutes. Beat eggs, milk and thyme together. Mix in the dulse and onion mixture and pour into the prepared pastry case. Bake at 350°F until set (about 20 minutes).

Black-a-Danny (Cornish dish, eaten as a main meal)
½ cup dulse
450 g (1 lb) blackberries
sponge pudding mixture for a steamed pudding
knob of butter
3 teaspoonfuls thick honey

Freshen dulse in water for 2 minutes, then drain. Steam blackberries in a basin until the juice runs, then squash the fruits. Stir honey into blackberry purée. Sauté dulse in the butter for 5 minutes, then stir into the blackberry mixture. Make up the sponge pudding mix a little on the dry side, so that it does not become too wet when the blackberry and dulse mix is stirred in. Put into a basin with a cloth tied over and steam for 3 hours. Serve hot with thin cream.

Dulse soup
450 g (1 lb) fresh wet dulse
1.25 litres (2¼ pints) water or clear stock
75 g (3 oz) oatmeal
seasoning

Rinse the dulse in clean water, then bring to boil in water or stock. Simmer for about an hour. Strain out the weed. Keep the remaining stock. Make up oatmeal to a thin cream with a little water. Add to the stock carefully, stirring all the time. Simmer for 10 more minutes, then season to taste.

Do not throw away the dulse particles strained from the soup. Mix with oatmeal, pepper, a bit of fat and perhaps a pinch of herbs. Fry in pats and use as a vegetable.

Porphyra umbilicalis
The traditional sloak or laver, eaten in South Wales, North Cornwall, Scotland and Ireland. In Japan it is the cultivated nori, a favourite food with everyone. There are many species of *Porphyra*; *P. umbilicalis* is one of the coarser varieties, but most common; other species are thought to have a better flavour and be more tender.

Laver bread
2 cups laver, fresh
1 cup rolled oats or oatmeal
bacon fat, butter or margarine to fry

Spread cleaned fronds on a plate sprinkled with oatmeal. Put more oatmeal and fronds on top, finishing with a layer of oatmeal. Roll the leaves into a wad, slice through with a very sharp knife. Fry the slices.

Other uses
Laver is such a versatile vegetable that a whole book could be dedicated to laver recipes alone. You are recommended to dip into Judith Cooper Madlener's book (*see* Bibliography) for Japanese specialities. Meanwhile, to whet your appetite, here are a few things to do with laver:
Chop fresh laver into a wild herb salad (goes well with angelica stalk).
Poach fronds in a little water, with a dash of cider. Serve as a vegetable.
Brush fresh fronds with oil and salt, stack and leave for 10 minutes, then toast in a heavy and hot frying pan till crisp. Use with dips.
Chopped fresh weed, mixed with egg and breadcrumbs, makes mock fish fingers.
Throw fronds into soup.
Roll fronds into balls and put in stews.

IMPORTED DRIED SEAWEEDS
There are two major suppliers of dried seaweeds in this country. One is Sunwheel Foods of London, the other is Westbrae Natural Foods, USA. Both market Japanese products, and in addition Sunwheel sell British specialities such as Irish moss and dulse.

All the packets of dried weeds carry basic recipes, so I will not repeat them here. Also an excellent book, *The Sea Vegetable Book*, by Judith Cooper Madlener gives many Japanese and other foreign recipes, which can only be tasted in the UK if you purchase the imported seaweeds. Most varieties of the popular Japanese seaweeds can be found in British shops, though you may have to shop around, or persuade your local friendly wholefood store to order specially.

Spirulina
A primitive single-celled alga that is found on the surface of alkaline lakes from Africa to Mexico. It belongs to the family of blue-greens or *Cyanophyta*, and under a microscope looks like beautifully coiled blue springs. It is the richest source of protein

Red seaweed	American name	Japanese name	Forms available
Bangia fuscopurpurea	hair plant	ushi-ke-nori	a delicacy
Gloiopeltis furcata	gumweed	funori	fresh or pickled only
Green seaweed			
Monostroma latissimum	sea vegetable	awo-nori	dried sheets
Brown seaweed			
Analipus japonicus	sea fir-needle	matsuma	dried
Eisenia bicyclis		arame	sweet, nutty. Dried
Hizikia fusiforme	sea weed	hijiki	dried, popular
Kjellmaniella gyrata		tororo kombu	dried
Laminaria angustata		mizu kombu	dried

Laminaria of all sorts go under the name of Kombu and are imported dried. (With so much of our own free for the gathering, it seems pointless to pay for high priced imports.)

	American name	Japanese name	Forms available
Nemacystus decipiens		mozuku	salted or brined
Undaria pinnatifida		wakame	dried, very popular

yet discovered, high in the B vitamins and vitamin A. If taken on an empty stomach, *Spirulina* makes you feel well fed, and the goodies contained in it ensure that your body receives nourishment, without the calories. No wonder it is being mooted as the slimmer's answer to a prayer. Because *Spirulina* fixes its nitrogen straight from the atmosphere, a bit of intensive cultivation could make one small lake the most productive 'farm' on earth.

Spirulina powder can be purchased at health food stores. Mix it with flour when baking. Put a teaspoonful into stews and soups.

Spirulina booster for athletes
1 tablespoon runny honey
1 tablespoon *Spirulina* powder
0.5 litre (1/3 pint) skimmed milk
2 teaspoons lecithin powder
0.5 litre (1/3 pint) pure orange juice

Blend all ingredients well and chill slightly. Take 4 kelp tablets with each drink.
N.B. *Do not* put raw egg in this drink, as is popular. Raw eggs prevents proper absorption of B vitamins by the body.

Agar-Agar

Of all the seaweed products, this is the one eaten by most people, often unknowingly. The multitude of commercial uses will be mentioned later (see p. 154), but even in its undisguised form it is a very useful substance to have around the kitchen. Although by itself it is tasteless and with virtually zero calories (a useful item for slimmers), it seems to accentuate the natural flavours of other foods. Agar jelly will set at room temperature, but not if acetic acid, e.g. wine vinegars, or oxalic acid items such as chocolate, spinach and rhubarb, have been mixed in. Lemon juice, malt and cider vinegars do not stop the gelling action.

Powdered agar can be bought at most wholefood stores and chemists in this country, and keeps indefinitely in an airtight jar. You can try and make your own, but the results will vary from moderate to disastrous, as so much depends on which species of red seaweed you picked, the season, what the weather had been like the week before, and probably the phase of the moon as well. The best agar comes from species of *Gelidium*, but you could pick other reds such as *Pterocladia, Gigartina, Ahnfeltia, Gracilaria* and *Acanthopeltis*. Try to gather the weeds after a spell of sunny weather from May to July. Wash and spread the plants out to dry. Boil the dry material in a solution of KOH (Potassium hydroxide): just two drops to a pint of water. KOH is commonly used in making soap, and your friendly neighbourhood chemist should have a drop to spare. Rinse the weed and then boil up in fresh water containing a generous squirt of lemon juice. Don't use aluminium pans for this operation, or enamel if it is chipped. The very best is a stainless steel pressure cooker. Cook for at least one hour, then throw in a handful of edible chalk powder. All the agar should curdle and sink to the bottom of the pan. Now put the whole lot in the freezer. When you are ready to play with it again let it defrost, add fresh water and drain through a jelly bag, repeating this freeze, thaw, wash system until you think it looks right. Then spread out onto a tray to dry. It should now be agar, but only you will know. Test by dissolving a little in hot water and see if it sets to a jelly.

Commercially produced pale agar is tasteless, and one level teaspoonful should set half a pint of liquid. Just sprinkle the powder onto the boiling liquid and stir until dissolved. If you purchase a darker powder 'Gelozone', then dissolve it with a little cold water first, and pour the hot liquid over. Gelozone has a strong flavour and does not set as a clear jelly.

The following recipes give just an indication of the versatility of agar.

Green soup (a slimmer's delight)
½ teaspoonful Gelozone
1 large leek
¼ small white cabbage
1 clove garlic
bay leaf
1 1/3 litres (2 pints) liquid
2 sticks celery
1 large onion
sprig parsley
a touch of salt and pepper

Chop up onion and garlic very small; put into pan (preferably stainless steel, with well fitting lid) with the bay leaf and add a little water. Simmer until water has almost gone and the onion is soft. Remove the bay leaf. Chop up the greenstuff and add to onion mixture. Put in rest of water and bring to boil. Mix Gelozone with a little cold water, then add to boiling vegetables. Remove from heat. Vegetables should be bright green and crunchy, in a slightly thickened flavoursome liquid. The Gelozone provides vitamins, minerals and protein, making this a good nourishing meal.

Melon shrimp cocktail
1 level teaspoonful agar-agar
1 ripe melon
small tin shrimps or prawns
2 sprigs watercress
1/3 litre (½ pint) liquid (including liquor from tinned fish)
½ teaspoon light honey
fresh ground black pepper

Halve melon and remove seeds. Scoop out about half of the flesh. Bring liquid to boil and sprinkle in agar, stirring until dissolved. Remove from heat, add honey and a twirl of freshly grated black pepper. In the halved melons arrange layers of shrimp, chopped watercress and slivers of melon. Carefully spoon in the jelly mixture until level with the cut top of the melon. Leave in a cool place to set. To serve, slice into segments and display on a bed of green lettuce.

Rainbow mould

Although this is a savoury mould, it could just as easily be a sweet version with multi-coloured fruits.

3 level teaspoonfuls agar-agar
575 ml (1 pint) water
assorted vegetables chosen for their colours: red and green peppers, grated carrot, yellow tomatoes, red cabbage (not pickled), pale green chicory tips, etc (approx 2 kilos)
egg yolks

Bring water to boil and dissolve agar-agar. Remove from heat and put aside to cool while vegetables are prepared. If jelly sets it can be re-liquidized by gentle heating. Arrange the finely chopped vegetables and the egg yolks in a rinsed mould, fixing each layer down with a little jelly spooned carefully over, and allowing to set before the next layer is arranged. Start with dark green, then light green, yellow, orange, red and purple, to give a rainbow effect. Turn out onto a bed of watercress, and serve with fried noodles, crunchy bean sprouts, pecan nuts and soy sauce.

Orange sparkle

2 level teaspoons agar-agar
2 tablespoons honey
satsuma or mandarin segments, chopped
290 ml (½ pint) water
290 ml (½ pint) orange juice
1 teaspoonful lemon juice
whipped cream to decorate

Put water and honey into pan and heat carefully until boiling. Sprinkle agar onto liquid and stir until dissolved. Remove from heat. Add orange and lemon juice, mixing well. Arrange fruit in bottom of glasses and spoon a little of the jelly mixture in to 'stick' the fruit to the glass. Pour in rest of jelly and leave to set. Decorate with whipped cream. For a special occasion dip the rim of a glass into the jelly mix and then into coloured coffee sugar crystals.

Mock caviare

½ teaspoonful agar-agar or Gelozone
½ kilo (1 lb) fish roe, the granular sort
290 ml (½ pint) water
soy sauce

Boil water and dissolve the agar. Set aside to cool. Mash the fish roe with soy sauce to give a nice rich golden colour. Stir in jelly liquid until it forms a thick mass that slumps slowly from the fork. Spread as a layer about ¼ inch thick over the bottom of a tray and leave to set firm. Serve either cut into fingers, or remashed, on tiny fingers of toast.

Fish paste
½ teaspoon agar-agar
any fish bits (pilchards are excellent)
290 ml (½ pint) water
tomato purée to colour
dash of anchovy sauce

Boil water and dissolve agar. Set aside to cool. Mash fish with tomato purée and the dash of anchovy sauce. Beat in jelly mix and leave to set. Mash again. Spread on fingers of toast or fill vol-au-vent cases. A larger quantity could be set in a fish mould as the starter to a meal, served with salad and buttered brown bread.

Additives
Processed seaweed additives to foods have now been given E numbers. (E numbers are additives that are generally recognized as safe under European Common Market law. They must be included in the list of ingredients on all food packages.)

E400 Alginic Acid: A natural product with no known adverse effects. From this are derived the following alginates:

E401 Sodium alginate

E402 Potassium alginate

E403 Ammonium alginate

E404 Calcium alginate (Algin)

E405 Propane -1, 2-diol alginate (Propylene glycol alginate, alginate ester)

All have no known adverse or toxological effects.

E406 Agar (agar-agar, Japanese isinglass), no known adverse effects in normal consumption quantities.

E407 Carragheenan (Irish moss). In 1981 a report in the *Lancet* (February 7th, page 338) stated that possible adverse effects had been found. Carragheenan may be one of the causes of ulcerative colitis, and possibly carcinogenic when broken down (degraded). The most harmful form is when taken in a drink. This needs further research because such adverse effects have not been found among people who use the whole plant and process it themselves. It could be another case of the purified form being 'too good to be true'.

9

Pollution – Paradox and Challenge

Why a chapter on pollution in a book that is mainly about food? It is a subject that touches us all, every day, but we become particularly aware of it as a problem when we decide to gather our own food from the wild. A simple activity that people used to pursue as an essential part of daily life is now fraught with problems and hidden dangers. Would you pick and eat herbs grown in the central reservation of a motorway? No, of course not, you can see and smell the danger from car fumes, dust, dirt and oil. Would you eat mussels grown in a radioactive river? How would you know the difference, for despite popular jokes, they do not glow in the dark.

Humans have been using the sea as a refuse tip since man first walked the earth, and the waters obligingly covered up and carried away anything left below high tide mark. Sea creatures disposed of the organic waste, waves pounded up and ground down anything too bulky, then sand and silt buried the evidence. From ships to sewage there were few traces left except to the eagle eye of the marine archaeologist. Old-style pollution by ordinary human waste was an acceptable form of nutrient recycling. But there were not so many humans around then. Nowadays it is different because people have disturbed the cycle.

There are two types of pollution in the sea: self-pollution and man-made. The first is caused when there is a natural outbreak, either from the planet itself as in a volcanic eruption underwater, or in a biological bloom. Sometimes something triggers a massive population explosion in a particular organism, usually one of the plankton. Rapid reproduction of one species takes place at the expense of the others around it. All available food, solid or liquid, is used up, and the waste products can poison the immediate waters for other plants or animals. In addition the available oxygen may also be used up, suffocating any opposition. These outbreaks are called 'red tides', because the sea often appears to be stained by organisms to a red or brown colour. Not only the sea but also rivers and lakes can show these phenomena; remember the river of blood recorded in the Old Testament, a classic case with fish dead and insects proliferating when their predators had been wiped out. Such phenomena are short-lived, and the waters recover eventually, creatures moving in from outside the damaged area until everything is back to normal. Man-made pollution is far more insidious.

A mile from our village there is a pretty little cove, typical of the south Cornish coast. A small settlement of fishermen's cottages and newer bungalows look down on a curved beach guarded by jagged rocks, backed by a steep cliff. In the summer tourists mingle with seabirds on the rocks and in the pools, turning over stones and seaweed to find crabs and any other small sea creatures. Pollution is an emotive word, usually associated with some single spectacular event such as an oil slick. But a longer look at our pretty Cornish bay will show the real pollution problems. Storms bring up vast quantities of cast seaweed at various times of the year. In the 1940s this weed was collected and spread on the adjacent farmlands. Local records and memory say the cast every winter was four or five feet thick on the beach. Now, in the 1980s, it is no more than a foot (0.3 metres) deep, small weeds mixed with rags, plastic, dead oiled seabirds, tar, rope, and discarded bits of fishing nets. In the water, invisible solutions of heavy metals and radioactive waste from the nearby naval dockyard mix with the raw sewage output of the bungalows and the run-off of chemical fertilizer overdoses from the farms. Occasionally there is a local 'scare' as drums of something toxic wash ashore, dumped by passing ships. The odd unexploded shell is found by visitors padding near the sewer outfall pipe while collecting shellfish for tea. After an oil spill the beach is yellow with the pretty little shells of winkles, killed, not by the oil but by the detergent sprayed on to break up the slick. Locals collect the shells and turn them into jewellery to sell to the tourists. In the aftermath the water is beautifully clear at a time when the spring plankton bloom should be in full spring. Even the plankton are dead, those same microscopic plants that produce the very oxygen we breathe.

A grim picture but one that is repeated in many places all over the world. People take the oceans for granted, looking on them as bottomless pits in which to dump their rubbish while they play along their edges. But monsters tend to live in bottomless pits, waiting for the day that something awakens them . . . All the pollutants we throw into the waters seem to vanish, leaving little or no trace behind, lulling us into a false sense of security. Everything in the pond must be all right. The oceans perform a wonderful balancing act, keeping the accounts straight; diluting concentrated nasties, putting nutrients back into the food chain. But no one told the ocean that some nutrients are not for recycling, so mercury accumulates in tuna, heavy metals are gathered by shellfish, and radioactivity in seaweeds near the outfalls of nuclear power stations matches the levels in the surrounding seawater. Like any good investment bank, the ocean makes sure we eventually get back with interest everything we put in.

This balance is a paradox. In moderate amounts, even the most toxic of chemical inputs can be diluted and distributed with only minor ecological damage near the input site. A total kill in one area will be replaced with recruits from outside once the killer has gone. So scars in the habitat heal over quickly, and we think all is right with the world again.

Seaweed concentrates radioactive substances in itself where these are present in the water. But even off Sellafield nuclear reprocessing plant you would have to eat a vast amount of seaweed before your level of radioactive intake became of serious concern. Even then, the alginic acid in the seaweed would prevent you from absorbing the active metals into your body. A marvellous paradox, but typical of the oceans' great design. In a paper presented to the 7th Seaweed Symposium (see Bibliography) Yukio Tanaka showed that radioactive strontium in the gut is bound by alginate, turning it into an insoluble salt. This stops the body absorbing the strontium and the salt is excreted. As mentioned in the previous chapter on health and beauty, radio-strontium already in the body's tissues and bone is drawn out as well. In his book *Pesticides and Pollution* (Collins, London, 1967), Kenneth Mellanby notes that even if you consumed laver bread every day made from weeds near the Sellafield outfall, you would not increase your radioactive intake to much more than one tenth of the natural background levels. Whether this increase in radioactivity would be harmful is a matter for continued argument, and you are not recommended to go picking seaweed from the outfall area just to prove a point.

Algae will not grow in polluted waters. Japan found this out to her cost in the Seto Inland Sea, when the nori crop nearly died out. But pollution has a far reaching effect even in the open ocean. Many things depend on the algae for their life. At the bottom of every food chain you will find a plant, and the sea is no different to the land in this. For example, planktonic algae are eaten by smelt, which are eaten by cod, which are eaten by people, among other predators. Iguanas of the Galapagos Islands are direct browsers on the seaweeds. Even the dumping of something not in itself toxic, such as large quantities of sediment, can change an environment so much that seaweeds no longer grow. Large algae, the familiar weeds, need rocks to grow on, and clean water. Too much sand will cover up the stones and suffocate the plants, then the shellfish move away and the fish swim to better feeding grounds.

The ability of seaweeds to concentrate whatever is in the seawater around them has been researched by various scientists. In Poland, Czapke looked at the levels of radioactivity in the weeds after he put radio-tracers into the water. Metal uptake by *Fucus vesiculosus* in the Bristol Channel was studied by Morris and Bale, in 1975. From works such as these it is hoped that a method of quickly monitoring the state of the seawater in any one place can be devised. When the radiation goes the weed returns to normal levels very quickly, so could give almost instant and very cheap local environment information. With metals it is a little different. Some are taken up by the seaweeds in a passive manner, others seem to be controlled by the plant itself, so more work needs to be done on this subject to define the rates and limits of metal intake.

A report came out in 1983 called *The Conservation and Development Programme for the U.K.*, which raised the problem of long-term pollution and the lack of knowledge about possible side effects. The sea self-cleanses by locking persistent contaminants away in the sediment, but a change in sea bed pattern by storm erosion or earth movements could release those pollutants into the water again, over a very short period of time.

Sediments underwater and sand dunes on land have much in common, including constant movement in the direction of the prevailing current or breeze. On land, grasses are used to hold the surface of the dunes still; in the sea, the seaweeds can serve the same purpose. Several scientists, in particular Dr Jolliffe (Bedford College, London and previously Hydraulic Research Station, Wallingford) have advocated controlling sea bed erosions in coastal areas by 'planting' macro-algae.

What of the future? Everything said so far makes it look

gloomy, but change will come about if people want it. As an individual, care about where your rubbish goes. Start on the green path, others will follow. The thin end of the wedge of change went in years ago, now we begin to see awareness spreading: better food, care for the environment, so eat seaweed instead of sheep. Like bricks in a wall, each person's contribution matters to the integrity of the whole.

10

Commercial Uses:
Past, Present and Future

Commercial exploitation of seaweed and its products has been around for a long time, especially in Japan. But the vast reserves of plants fringing the world's oceans are still largely untapped because of the problems of gathering wild weed. Assessments made of seaweed resources around the Scottish coast just after the war estimated that over one million tons of weed could be taken from the coasts each year without depleting the supply. Nowadays, Scotland imports most of its seaweed from Ireland and other places. Japan is the world leader both in utilizing the natural resources of the ocean and in cultivating preferred species. Actual figures available for the worldwide commercial usage of marine plants do not mean anything, because politics tends to get in the way of proper data publication.

Originally in western countries seaweed was burned to produce soda and potash for use in the manufacture of glass and soap. Modern chemistry and the discovery of mineral potash deposits on land replaced the old ways, so the seaweed users turned to iodine production. This prospered for a while but eventually the chemists caught up once more, and the iodine industry died out. Many towns bear relics of this weed trade in their place and street names, for example Saltash in Cornwall, where they used to produce salt from seaweed, then potash and iodine, for use in the naval dockyards of Plymouth.

During the Second World War, ways of producing various substances from the wet weed were tried out in America and Canada. Heating dried weed in an iron retort gave off acetylene, heavy oils, ammonia, potash, bromine and iodine and the ash made good charcoal. After the war, discoveries of land-based deposits of raw materials that were cheap to mine, together with the increasing expertise of the industrial chemists, made processing seaweed for those same materials too expensive. The Russians developed an electrolysis system for extracting chlorine, bromine and iodine at the anode, cellulose and alginates at the cathode. Although there is at present little commercial significance in this process, it has great potential value should the country need to become self-sufficient in the products.

Because the harvesting of wild weed is so labour-intensive and the crop can vary so much in composition, various methods have been tried and used to cultivate seaweeds. In Japan, nets and bamboo blinds are made into artificial sea floors. Young sporling weeds attach themselves to the nets, which are then transferred to coastal sites where the water is rich in nutrients. These can be enriched where needed; for example, the Chinese release extra nutrients into the water from porous pots suspended over the nets. Monocultures of specific seaweeds are 'weeded' periodically to clear any other species, especially epiphytes (plants or animals that grow on other plants). The working of such seaweed gardens is still labour-intensive, but the crop is of greater purity and higher value. Principal sea vegetable crops in Japanese waters are *Porphyra spp* (nori), *Laminaria spp* (kombu) and *Undaria* (wakame).

Seaweeds usually dislike warm, low nutrient waters, and do not produce spores in these conditions. This has been overcome by having inland hatcheries. Once established the tiny plants are transported to the coast. With *Laminaria* (kelps) the plants grow in the best areas of water, with most food in their first years, then they are transplanted to other areas where they are allowed to grow on to maturity.

The USA is one of the few places where seaweed gathering has been mechanized successfully. Weed-cutting barges work the *Macrocystis* beds, cutting only the top metre of kelp. One barge can harvest more than thirty tons an hour. Despite government regulation on the depth of cut, the *Macrocystis* beds have been declining in productivity. Part of the problem may be pollution, but changes in hydrodynamics and local climate may also be playing their part. Various projects to increase the nutrient quality and quantity of the seawater in the weed beds have been tried with varying degrees of success. One imaginative scheme from Jackson and North in 1974 proposed bringing up artificially the cold rich waters from deep down to the depleted shallow warmer zone.

Because of the problems of gathering the wild weed harvest, or its open-water cultivated varieties, it was inevitable that attempts would be made to grow seaweeds in tanks on land where the processes of growth could be monitored and controlled. Harvesting would also be easier in such an environment. These projects have been surprisingly successful. Most seaweeds are notoriously difficult to keep 'in captivity' as any marine aquarium buff will tell you. The plants need nutrient-rich moving water and the right temperature at the right time for optimum growth. Give them these conditions and they grow happily. A production-line system would, ideally, be based in large plastic tanks in a long tunnel greenhouse, with pumped throughflow water system, nutrient monitoring, automatic temperature and feed controls. Original research in this type of tank cultivation was started by A.C. Neish in 1971. He used *Chondrus crispus* (carragheen) and not only grew the weed successfully, but also found a strain of *Chondrus* which he called T–4 that had fast growth and a high carragheenan content. This is one great advantage of growing monocultures of a seaweed species in tanks: individual plants can be monitored and the best encouraged to propagate. It is the same method that has been followed with land plants for centuries, resulting in the succulent and large size vegetables we eat today. Once they were all weeds.

But why this great interest in seaweeds, with money being put into research, especially in the Americas? We do not see fresh weed on the cold-slab next to the fish, or on sale alongside carrots and potatoes. Yet world production of weed is in thousands of tons. It all has to go somewhere, and a lot of it comes into the UK and Europe. The amount used in agriculture and horticulture is small, animal feed accounting for only a tiny fraction of the total imports. Seaweed meal and liquid seaweed fertilizers are not so widely used as their manufacturers would wish. There is, however, an upsurge in interest in organically grown foods which should put new life into the seaweed fertilizer industry.

The main use of seaweed is as processed products, carragheenan, agar and alginates, which go into just about everything you can think of. The following list is really only the tip of an incredible technological iceberg based on sea plant products.

Carragheenan
Provides that rich 'body' and thickness you expect in salad cream, cream cheese, instant desserts and sauce. It is a preservative for frozen fish, and clears the clouds from beer. The same jellyish properties make insecticides stick to plants, rubber have more bounce, and thickens out paper and textiles, giving a good smooth finish at the same time. It gives a nice gloss to leather,

smooth texture to toothpaste, and plumps up beauty creams. Special properties such as waterproofing and being fire retardant are used in the paint industry. 'Permitted emulsifier' on a label often means carragheenan.

Agar

Uses are interchangeable with carragheenan, but also used as the thickener in canned fish and meat. You can eat agar in everything from mayonnaise to icing. A large amount is used in the pharmaceutical industry, in initial research, where it is the plating medium for microbiological studies, and as the basis of capsules and potions for everything from dangerous drugs to laxatives. Agar can set solid enough for dentists to use it when taking impressions of teeth, yet it can be a super lubricant when tungsten wire is drawn out hot.

Alginates

Again, the uses are interchangeable with the other seaweed extracts, but the favourite one in America is to keep ice cream stable and smooth. At one time, nearly all the American output of alginates went into the ice cream industry. Other uses include cosmetics, plastics, the manufacture of soap and glass, fireproofing and waterproofing paper, fibreboard and textiles. From welding flux to lemon curd you will find the ubiquitous seaweed gel. The uses are seemingly endless, the list increases daily.

Most of this demand is still satisfied by gathering weed from the sea. Commercial growing of seaweeds around the coast of Europe, and especially the UK is hampered by the lack of legal protection for the waters. Coastal wetlands and salt-water lagoons are good places for weed cultivation but in many areas these are now valuable nature reserves, and large-scale seaweed culture would destabilize their delicately balanced ecosystems.

Another problem with seaweed production is that at the moment it costs a lot of money and energy to process wet weed into final product. However, scientists are researching this, and the ultimate answer may be to make the seaweeds pay for their own cultivation and conversion. Anaerobic (without oxygen) digestion of the plants can produce methane gas and fermentation gives ethanol (alcohol). Both these products can be used as fuel to run cultivation tanks, thereby preserving the wild seaweed population, and also to dry the harvested plants.

A side of seaweed cultivation rarely talked about is as a sewage treatment plant. Warm-water species of plants have been used successfully in a pilot project to design a one-step waste-recycling-aquaculture system, as reported by J.H. Ryther in 1976.

The future of seaweed looks bright: kelp may be used to produce petro-chemical analogues and industrial alcohol. This would be valuable as the world's oil resources begin to run out. Calcareous algae (*Lithothamnion*) is now used to neutralize acid drinking water, and prevent the corrosion of pipes. Perhaps in the next few years, tank-cultured *Lithothamnion* will provide the growing health food market with an 'organic' source of calcium and other minerals.

Back in 1956, Isaacs cultivated grown seaweeds from pieces of chopped frond. This led on to experiments in which a rope 4 metres long yielded a crop of 13 kg of weed in six months after being seeded with fragments. Now work is under way using tissue culture techniques to raise weeds from single cells. Future seaweed crops could be tanks full of clones, a single productive plant acting as the source for all the future generations: a marine version of the banana. (In case you didn't know, the banana does not produce seeds; all modern banana plants are daughters of the original mother plant.)

From a simple food to the most purified agar on a microbiologist's plate, seaweed is in great demand. Perhaps as sea products become more desirable, the threat of pollution will diminish as both commercial pressure and public awareness force the land-bound polluters to mend their ways.

Bibliography

General references
Barrett, J. and Yonge, C.M., *Collins Pocket Guide to the Seashore* (revised edition) (Collins, London, 1958)

Campbell, A.C., *The Hamlyn Guide to the Seashore and Shallow Seas of Britain and Europe* (Hamlyn, London, 1976)

Chapman, V.J., *Seaweeds and their Uses* (Methuen & Co. Ltd, 1970)

Chapman, V.J., *Coastal Vegetation* (2nd edition) (Pergamon, Oxford, 1976)

Dickinson, C.I., *British Seaweeds* (Kew Series, Eyre and Spottiswood, London, 1963)

Duddington, C.L., *Beginner's Guide to Seaweeds* (Pelham Books, London, 1971)

Hanssen, M., *Spirulina* (Thorsons Publishers, Wellingborough, 1982)

Hills, L.D., *Organic Gardening* (Soil Association Handbook, 1976)

Hunter, C., *Vitamins, What They Are and Why We Need Them* (Thorsons Publishers, Wellingborough, 1978)

Lovelock, J.E., *Gaia: A New Look at Life on Earth* (Oxford University Press, Oxford, 1982)

Madlener, J.C., *The Sea Vegetable Book* (Clarkson N. Potter, New York, 1977)

Major, A., *The Book of Seaweeds* (Gordon and Cremonesi, 1977)

Mellanby, K., *Pesticides and Pollution* (Collins, London, 1967)

Myers, N., *The Gaia Atlas of Planet Management* (Pan, London, 1985)

McDonald, K., *Food from the Seashore* (Pelham Books, London, 1980)

Polunin, M., *Minerals, What They Are and Why We Need Them* (Thorsons Publishers, Wellingborough, 1979)

Rhoads, S.A., *Cooking with Sea Vegetables* (Autumn Press, Massachusetts, 1978)

Stephenson, W.A., *Seaweed in Agriculture and Horticulture* (Bargyla and Gylver Rateaver eds (Conservation Gardening and Farming Series, California, 1974)

Wilson, F., *Kelp for Better Health and Vitality* (Thorsons Publishers, Wellingborough, 1979)

Specialist references
This list is meant only as an introduction to the vast amount of literature available on the subject of seaweed.

Boney, A.D., *Biology of Marine Algae* (Hutchinson, London, 1966)

Chapman, V.J., *The Algae* (Macmillan, London, 1962)

Cook, G.W., *Fertilizing for Maximum Yield* (2nd edition) (Granada Publishing, London, 1976)

Fortes, M.D. and Luning, K., 1980. 'Growth rates of North Sea macroalgae in Relation to Temperature, Irradiance and Photoperiod', *Helgolander Meeresunters.*, 34: 15–29

Foster, P., 1976. 'Concentrations and concentration factors of heavy metals in brown algae', *Environmental Pollution* 10: 45–53

Nebb, H. and Jensen, A., *Proceedings of the Fifth International Seaweed Symposium* (Pergamon Press, Oxford and New York, 1966)

Soil Science, 7th International Congress, 1960, IV, Baltimore, USA

There are many such congress reports on seaweeds, all worth looking up.
Postland, I. chairman. 'The Conservation and Development Programme for the U.K. – A Response to the World Conservation Strategy.' 1938. World Wildlife Fund
Smith, A.M., *Manures and Fertilizers* (Nelson's Agricultural Series, Thos. Nelson and Sons, 1970)

Suppliers
Dickins and Jones, Regent Street, London W1A 9DM. For seaweed massage cream, soap and cream bath.

Lincoln Fraser Products, Chiltern House, Gold Hill West, Chalfont St Peter, Bucks. SL0 9HH. For Aura Som seaweed mineral bath.

Joy Byrne, 37 Albemarle Street, London W1, uses seaweed products in her beauty treatment centre.

Creightons seaweed shampoo available in good chemists.

The Body Shop for seaweed and birch shampoo.

Foods
Westbrae Natural Foods, Berkeley, CA 94706, USA

Sunwheel Foods Ltd, Orpheus Street, London SE5 8RR

Index

158